Radio's Digital Dilemma

Radio's Digital Dilemma is the first comprehensive analysis of the United States' digital radio transition, chronicling the technological and policy development of the HD Radio broadcast standard. A story laced with anxiety, ignorance, and hubris, the evolution of HD Radio pitted the nation's largest commercial and public broadcasters against the rest of the radio industry and the listening public in a pitched battle over defining the digital future of the medium. The Federal Communications Commission has elected to put its faith in "marketplace forces" to govern radio's digital transition, but this has not been a winning strategy: a dozen years from its rollout, the state of HD Radio is one of dangerous malaise, especially as newer digital audio distribution technologies fundamentally redefine the public identity of "radio" itself.

Ultimately, *Radio's Digital Dilemma* is a cautionary tale about the overarching influence of economics on contemporary media policymaking, to the detriment of notions such as public ownership and access to the airwaves—and a call for media scholars and reformers to engage in the continuing struggle of radio's digital transition in hopes of reclaiming these important principles.

> Anderson provides a detailed and shocking look into how compliant regulators and a few well-connected private actors can conspire to thwart both the market and the public interest. This is a startling and well-documented indictment of an epic failure of our media system that should enrage both liberals and conservatives alike.
>
> —*Ted M. Coopman, San Jose State University, USA*

Anderson's text elucidates an important, and overlooked, policy fight. Largely out of sight of the public, and over an extended period of time, broadcast conglomerates and related interests pushed to replace our current open broadcast system with one based on a technically deficient, proprietary standard, that they controlled. Anderson has forensically assembled this story, showing us how obscure policy battles over technical standards can have long-reaching impacts on the media that act as conduits for so much of our culture.

—*Andrew Ó Baoill, Cazenovia College, USA*

John Nathan Anderson is an Assistant Professor and the Director of Broadcast Journalism in the Department of Television and Radio at Brooklyn College, City University of New York. Formerly a radio journalist, he's been working in the fields of media policy and activism for nearly two decades.

Routledge Research in Cultural and Media Studies

For a full list of titles in this series, please visit www.routledge.com

Radio's Digital Dilemma

Broadcasting in the Twenty-First Century

John Nathan Anderson

Routledge
Taylor & Francis Group

NEW YORK AND LONDON

First published 2014
by Routledge
711 Third Avenue, New York, NY 10017

and by Routledge
2 Park Square, Milton Park, Abingdon, Oxfordshire OX14 4RN

First issued in paperback 2016

*Routledge is an imprint of the Taylor & Francis Group,
an informa business*

Library of Congress Cataloging-in-Publication Data

Anderson, John Nathan, 1973–
   Radio's digital dilemma : broadcasting in the twenty-first century / by John Nathan Anderson.
      pages cm — (Routledge research in cultural and media studies ; 60)
   Includes bibliographical references and index.
 1. Digital audio broadcasting.   I. Title.
   TK6562.D54A53 2014
   384.54'52—dc23
   2013031862

ISBN 13: 978-1-138-65149-4 (pbk)
ISBN 13: 978-0-415-65612-2 (hbk)

Typeset in Sabon
by Apex CoVantage, LLC

Contents

Acknowledgments

Putting this book together has been, for the most part, an intensely monastic experience. But there are many people who were instrumental in helping me shape and complete this project—whether it was through thought-provoking discussion, sage advice about the research and writing process, review and critique of chapter drafts, and/or generally helping me maintain a semblance of sanity along the journey: Marko Ala-Fossi, Ann Alquist, Nate and Betsy Anderson, Ryan Atteberry, Susan Brandscheid, Tammy Brandt, Frank Cerciello, Eric Cobb, Ted Coopman, Pilar De Silva, Liz DiNovella, Brian Dolber, Brian Dunphy, Jay Eychaner, Jennifer Fandel, Denise File, Edwin Hadley, Tim Hansel, Todd Hunter, Angharad Johns, Brad Johnson, Kristian Knutsen, Ellen Knutson, Benn Kobb, Jeff Kolar, Kamilla Kovacs, Matthew Lasar, Bob McChesney, Lynsee Melchi, Tim Meyers, Chris Miller, Craig Mingus, Michelle Nelson, John Nerone, Andrew Ó Baoill, JoAnne Powers, Paul Riismandel, Felisa Salvago-Keyes, Christian Sandvig, Thomas Scaggs, Zack Steigler, Molly Stentz, Norm Stockwell, Christopher Terry, Angharad Valdivia, John Wason, Dylan Wrynn, Desiree Yomtoob, and Enid Zafran. Special thanks to Zuzu and Patience for their unconditional love.

I am also grateful to the Brooklyn College Department of Television and Radio for providing research assistance. Additional support for this project was provided by a PSC-CUNY Research Award, jointly funded by the Professional Staff Congress and the City University of New York.

1   Identifying Radio's Digital Dilemma

Radio broadcasting is the last bastion of the traditional analog mass media to negotiate the communicative phenomenon known as convergence—the ongoing evolution of media technologies toward a universal digital communications language and platform, presently best exemplified by the Internet.[1] Convergence is itself a phenomenon governed by three factors: the development of new technologies, industry strategy, and public policy.[2] Although most analyses of convergence focus primarily on its technological aspects, the phenomenon is more often than not "the product of political will, rather than inexorable logic."[3] Those involved in the crafting of communications policy often promise us that new media technologies will make our media environment fundamentally more democratic—but in many respects, convergence has opened up the potential for potent unsettlement, especially with regard to how the phenomenon shapes legacy media systems.[4] Unfortunately, corporate interests have skewed the regulatory development of our convergent media environment to entrench the priorities of commerce above all others; from the perspectives of industry strategy and public policy, convergence is a convenient vehicle by which to bring the "entire 'ideological' sphere of society" into the orbit of a hyper-capitalist political economy.[5]

Changes taking place in the realm of radio broadcasting are quite illustrative regarding the perversely negative effects that convergence can have on a medium undergoing a digital evolution. On one hand, as discussion of media digitalization gained steam during the 1990s and the commercialization of the Internet engendered the formal study of convergence itself, the U.S. radio industry—and the Federal Communications Commission that oversees it—remained surprisingly insular in their outlook for the medium's future. The 1990s were a decade of consolidation and the reconfiguration of radio relative to its mission to serve "the public interest, convenience, and necessity" into a fully entrenched marketplace paradigm. In simple terms, many involved in radio broadcasting have conceptualized convergence as a tool that allows the programming of multiple stations from one location and has been utilized more for cost-savings than for the creative or communicative expansion of broadcasting itself. By the time regulators, media professionals, and scholars began to grapple with the implications of convergence, the radio industry as a whole wasn't even fully aware that the phenomenon was underway, much less what its implications for the medium itself might be.[6]

The sheer expansiveness of the Internet and the introduction of digital satellite radio broadcasting served to change this ignorance, and the need for radio to address convergence became a growing topic of debate within the U.S. broadcast industry around the turn of the twenty-first century. However, broadcasters initially defined digitalization *as* convergence, when in reality digitalization is only the first step in navigating the phenomenon. To be sure, it is an important step, and David Sedman has identified four criteria necessary for the adoption of any new radio service: "(1) Approval by a governing body (such as the FCC in the United States); (2) Acceptance by the broadcast station; (3) Consent from the consumer electronics industry to design and market a new technology; [and] (4) Adoption by the mass buying public."[7]

Unfortunately, the technology developed by the U.S. radio industry and sanctioned by the FCC actually represents one of the worst possible iterations of digital radio, and does more to fragment the medium's participation in the convergence phenomenon than it does to embrace it. Only one of Sedman's four criteria (regulatory approval) has been met, primarily due to the fact that the FCC didn't even bother to examine the real-world implications of radio's digitalization. Instead, the nation's most economically powerful broadcast incumbents sold radio's digital transition as an evolutionary necessity, trusting in the "market exchange ethic" of neoliberal ideology to govern all aspects of digital radio's development and proliferation.[8] Thus, the policy and technology of digital radio in the United States were essentially crafted to serve selfish needs that fall far short of serving the public interest.

Radio's digital dilemma ultimately arises from a conflict between the traditional strengths of the medium and the perceived interests of broadcasters badly navigating a convergent media environment, enabled by a captive regulator. This has led to a state of affairs whereby radio's digital transition is effectively marginalizing traditional broadcasters; coupled with the rise of new digital audio delivery systems, the public identity of "radio" itself is now in flux.[9] This book attempts to explain the conditions that led to the present state of affairs. I argue that the political and economic decisions locking terrestrial broadcasters into a questionably viable digital radio technology has dangerous implications for the integrity of the medium as we've known it,[10] and I aim to illustrate how the policymaking process—which is ostensibly designed to maximize the functional utility of media systems—has been effectively privatized in the case of radio, leading to such a potentially deleterious outcome. The most important aspect of this story is not the technology itself under scrutiny, but the values it embodies and their implications for the future of broadcasting.

DIGITAL RADIO AS A CAPSTONE IN NEOLIBERAL MEDIA REGULATION

In the United States, corporations directly control the development of most new media technologies, which are then deployed with little room for critiques of the political-economic structure that surrounds them.[11] Within the regulatory process itself, the "sovereignty of citizens over the state" has effectively disappeared; in effect, the "citizen" has been replaced by the "consumer" in regulatory discourse more broadly.[12] In the case of the U.S. digital radio transition, constitutive choices were centered on the primary goal of broadcast incumbents to use technology as blocking mechanism to prevent new competition in the radio broadcast space.[13] In fact, the story of radio's digital transition may be one of the clearest examples of how a stringently neoliberal ideology can destroy democratic media systems as they fall under a private paradigm of control, where even regulators cannot intervene.

Paul Starr first articulated the concept of constitutive choices: they represent decisions made during the creation of media systems that affect "how things are built and how they work—their design and rules of operation." Constitutive choices are often made through a process of "slowly crystallizing cultural practices or gradual economic and political change," but in some cases they arise in "bursts set off by social and political crises, technological innovation, or other triggering events, and at these pivotal moments the choices may be encoded in law, etched into technologies, or otherwise embedded in the structure of institutions."[14]

Constitutive choices have three primary effects on the nature of democratic communications: they determine, first, "the general legal and normative rules concerning such issues as free expression, access to information, privacy, and intellectual property; second, the specific design of communications media, structure of networks, and organization of industries; and third, institutions related to the creation of intangible and human capital— that is, education, research, and innovation."[15] Good constitutive choices are made when all three factors are taken into account and balanced in such a manner as to maximize the democratic potentiality of a communicative system.

In the United States, the dominance of neoliberalism over all other ideologies in the process of radio regulation has deep roots. Susan Douglas, Mark Lloyd, Robert McChesney, Susan Smulyan, and Thomas Streeter all assert that critiques of capitalism were brushed aside during the formative years of radio policymaking,[16] even though this exclusion was openly contested along the way.[17] The history of radio broadcasting itself, as McChesney puts it, "has the earmarks of a history written by the victors," as it casts the development of contemporary broadcast regulation as a natural process, when it most definitely was not.[18] This ultimately leads to conditions by which bad constitutive choices get made. Starr himself argues that from the inception of broadcast regulation, commercial forces have worked "to create a set of rules . . . that virtually amounted to a system of private regulation of politics."[19] From the perspective of Thomas Streeter, the "key players" of radio broadcasting "have always been organized along corporate lines," which guarantees that an ideology of corporate rationalization "permeate[s] the institution of . . . broadcasting at numerous levels,"[20] and creates a stratification of power among constituents involved in policymaking—where the "core [is] dominated by an alliance of corporate and government elites," and a subaltern "economic periphery of smaller enterprises and a political periphery of electoral politics" exists.[21] It is, thus, all but impossible for U.S. media policymaking to claim that it serves any notion of democratic design:

> Broadcast policy is a realm for experts, not for "politics" in the broad sense of governance in a democratic society . . . And when those inside the delimited broadcast policy world knowingly acknowledge that policy is political, they mean political in the sense of maneuvering for gain—low politics. The world of policy, they readily acknowledge, has become infected by the processes associated with self-interested strategizing and struggles. But they don't describe the subject matter . . . as "broadcast politics" because this might imply high politics: matters of value, structure, and legitimacy that they and their sponsors have little interest in opening for consideration.[22]

If such a paradigm has dominated U.S. broadcast regulation from its inception, it becomes easier to understand how such poor constitutive choices for the digital future of radio are facilitated.[23]

The historical ascendancy of corporate control over radio policy dovetails well with the constitutive choices made about the medium's digital future. Douglas notes that the Radio Act of 1912 set a "watershed" precedent: "that only consolidated institutions . . . could anticipate, implement, and protect 'the people's' interest in spectrum use."[24] C. Twight implicates regulators with repeatedly and strategically distorting technical information on which the Radio Act of 1927 was constructed in order to pursue the implicit objectives of the nascent commercial radio industry.[25] The Federal Radio Commission's first major policy action, the imposition of General Order 40, fundamentally skewed use of the airwaves toward commercial interests.[26] When allocating FM spectrum more than a decade later, the FCC would couch the process in terms of technicalities, while obscuring the political–economic motives behind the action.[27] The Reagan-era FCC formally launched the neoliberal crusade to reclassify broadcasting as a commercial endeavor by dismantling many regulations that made broadcasters accountable to the public. More recently, the Telecommunications Act of 1996 "boldly equated the public interest with a competitive economic environment," put a heavy burden on regulators to justify their very existence, and encouraged them to pare their powers back whenever economic interests demand it.[28] McChesney now calls any idea of proactive broadcast regulation under current conditions a "living absurdity" and "tragicom-edy," given that notions of public service by broadcasters today are almost wholly defined by a corporate paradigm.[29]

The story of radio's digitalization in many ways represents the capstone of a communications policy trajectory defined by Nicholas Garnham as "a struggle to turn all information into private property and therefore a struggle of private profit rather than the claimed development of a system to provide information widely and cheaply to all."[30] In such an environment, the public is discouraged from participating in the policymaking process, is ignored if it does, and is generally notified of the outcomes after all constitutive choices have been made,[31] with casual disregard for normative concepts such as freedom of expression, public access to the airwaves, and the intellectual property aspects of broadcast technology.[32]

Institutionally, communications regulation in the United States occurs under burdensome constraints and pressure from several directions (industry constituents, Congress, and the courts). These impediments lead regulators to ignore or downplay actual public sentiment regarding the policies they promulgate—especially when the public attempts to assert a larger role in the policymaking process than regulators allow for. Mark J. Braun has catalogued the "immense workloads and small staffs" that "contribute to a hectic FCC decision-making environment,"[33] which only becomes more turgid when policymaking becomes publicly controversial. Walter B. Emery declared the Commission's resources "pathetically inadequate" more than 40 years ago, and the situation has only worsened over time.[34] The end result is policymaking done in an ad-hoc, typically conservative fashion, as the FCC attempts to placate all the groups that pressure it, rather than thinking proactively about the issue at hand.[35]

This has led to a situation which former Commissioner Nicholas Johnson once described as "a 'subgovernment' of industry lobbyists, specialty lawyers, trade associations, trade press, congressional subcommittee staff members, and commission personnel who dominate" the policymaking process.[36] The agency's practice of *ex parte* meetings, whereby constituents meet in person with regulators to discuss policy issues, further promotes regulatory capture. *Ex parte* conversations are a favored tool by which regulated industries manipulate the policy process to their own ends.[37] In a contemporary example, broadcast industry executives and lobbyists met behind closed doors with policymakers more than 200 times during the FCC's 2003 debate over the revision of media ownership rules; the agency ultimately attempted to promulgate policy in line with their desires, even though 97% of public comment received on the issue vehemently opposed this outcome.[38]

Regulatory capture is amplified when the FCC relies on the private sector for information that directly dictates the objectives of policy outcomes.[39] The FCC's lack of meaningful, independent analytical ability has been in clear decline for more than 30 years.[40] It thus comes as no surprise that if the data underlying a new policy favors a particular outcome, that outcome is effectively predetermined.[41] According to Philip Napoli, "On many points, commercial data collection and public policy needs fail to align. Commercial data is structured around the financial, investment, and marketing needs of media corporations and investors." Although the FCC conceivably "answers to a more complex concept of the public interest, which balances economic efficiency with concerns for equity, diversity, and constitutional rights,"[42] the end result of reliance on industry-sponsored analysis from which to craft communications policy means that "the communications industry's social and political functions are diminishing in importance relative to its economic function in the eyes of the FCC."[43] At the very moment when our information society requires both citizens and policymakers to be better versed in the intricacies of the technologies that define their lives,[44] the FCC appears to be moving toward what Sandra Braman characterizes as "narrative simplicity, even as the data upon which state narratives are placed become more diverse and complex."[45]

Within the agency itself, the makeup of key management staff has shifted away from people with legal or technical backgrounds to those with expertise in politics and economics.[46] This leads to the promulgation of regulation "without a sound empirical basis" that "contribute[s] to . . . inconsistency and ambiguity . . . in communications policy."[47] Former FCC Commissioner Jonathan Adelstein found the FCC's data collection process and decision cycle so devoid of actual facts that he dubbed the agency's activity "faith-based regulation."[48] Instances such as the FCC's 2003 media ownership rules review, according to Mark Cooper, were conducted without "any hint of intellectual or public policy integrity," and in recent years, when staff research has not meshed well with the political objectives of the FCC Chairman, it is suppressed outright.[49] Career civil servants at the FCC are very sensitive to the increasingly "arbitrary and capricious" nature of regulation in the corporate interest—so much so that on the third anniversary of Republican Chairman Kevin Martin's appointment in 2008, many came to work dressed in black in "silent protest" of the agency's politicization.[50]

So where, exactly, does the public fit in communications policymaking? Following the passage of the Communications Act of 1934, it was denied a formal voice in FCC proceedings until the imposition of the Fairness Doctrine in 1949 and subsequent judicial decisions that upheld the principle of public agency in the regulatory process.[51] Even so, there are many obstacles to obtaining information relevant to regulation, ranging from "opaque pricing structures and restrictive licenses for commercial data, to legal barriers to access, to the basic adversarial nature of contemporary communications policymaking" itself.[52] Thus, the public at large cannot afford to engage in the intensive level of discourse that the FCC deems necessary to be a true "player" in the policymaking process.[53] Public interest groups have repeatedly tried to intervene in policy debates, but the FCC has no coherent way of dealing with such input; the treatment of public comment is left up to the discretion of FCC staff working on a particular issue, and they are free to disregard it as they see fit.[54] Although electronic means of filing public comments now exist, FCC staff expects commenters to "do their home-work" and are much more inclined to disregard comments if they arrive in the form of petitions.[55] Non-governmental "media reform" organizations may make "the public" more visible in the eyes of regulators, but that is a far cry from policy outcomes that conform to any meaningful notion of the public interest.[56]

The end result is a democratically dysfunctional communications policy-making regime. The FCC's modus operandi often substitutes "the act of evaluating and studying a problem or policy for the act of actually dealing with a problem or making policy."[57] Braman comments that while policy-makers may *think* they are designing new regulations using the language of facts, "their role in policy-making is more likely to belong to the rhetoric of decision-making processes rather than their content."[58] Similarly, Philip Napoli has observed, "the foundational principles of communications pol-icy are used as rhetorical vessels to justify post hoc regulation, and typically do so from the point of view of dominant stakeholders in the process."[59]

Instead of being treated as a substantive issue that stands to radically transform radio, digital broadcast policy has been reduced to a marketplace paradigm heavy on metaphor but light on empiricism—ultimately driven not by regulators but by those whom they ostensibly oversee. In simple terms, a slow-motion coup has taken place over the course of 80 years of U.S. broadcast policymaking that now openly prioritizes private interests over the public interest. According to McChesney, if "the critical ques-tion facing us is whether the new technologies can rejuvenate . . . political democracy or whether the corporate, commercial domination of . . . the communication industries will be able to subsume the technologies within the profit net and assure that the corporate domination of both U.S. society and the global political economy remain unquestioned and unchallenged,"[60] the case of digital radio points dramatically in the latter direction. Although "critical scholarship remains a minority phenomenon"[61] and substantive debate "regarding the control and structure of the media" may be diffi-cult under such circumstances, when the future of radio broadcasting is at stake it is incumbent upon scholars to pay closer attention to the issue and learn lessons from the development of digital radio policy that may be applied to resist the neoliberal paradigm affecting our entire convergent media environment.[62]

THE DRAMATURGY OF TECHNOLOGY
AND POLICY DEVELOPMENT

A useful way to examine how such troublesome processes work involves employing a dramaturgical framework—assessing how the representa-tion of actors and arguments differs in certain key contexts, especially as it relates to disjunctures between words and deeds.[63] This allows for the careful examination of the actors behind these processes and how they manipulate discourse in order to achieve desired outcomes.[64] In the specific arena of media policy, Jan Ekecrantz believes dramaturgical analysis is effective in illuminating "intricate institutional and other power relations, which imbue them with meaning and constitute their *sine qua non* in the first place."[65] The regulatory analyses of Sandra Braman, Erwin Krasnow and Lawrence Longley, Philip Napoli, Dallas Smythe and Thomas Streeter imply that much of what passes for media policy is dramaturgical, for it allows the powerful to advance ideas that are given "credibility relative to their standing," as well as providing "a generalized immunity to perceptions of risk and danger that their activities might otherwise produce."[66] In the context of digital radio policymaking, these two points are critical to understanding the process that led to such a tenuous outcome for the future of the medium.[67]

Much of the dramaturgy conducted in modern society involves practices of communication that Peter K. Manning argues can be distortionary to core notions of "truth" on which the structure of sociopolitical power rests.[68] Manning also asserts, "The higher the trust in the industry, the lower the level of information required and produced by the industry."[69] In general, "powerful interests do not maintain their control as much by persuading us to believe them but, more often, by preventing us from knowing what they are doing."[70] Streeter notes that communications policymaking takes place in an interpretative community where shared meanings of certain ideas are taken for granted, and there is ample evidence that such behavior does not produce the most rational outcomes.[71] In this context, the venues in which policy-related discourse takes place become realms for "experts, not for 'politics' in the broad sense of governance in a modern neoliberal society"; fundamental matters such as "value, structure and legitimacy" thus become difficult to debate and democratically refine.[72]

Radio's digital transition is playing out on two primary dramaturgical stages. The first is the FCC's policymaking process, which represents the "official" discourse of the transition. Those involved in this process define their objectives primarily by the policy outcome they'd like to see. These objectives are not always clear, and the FCC's practice of *ex parte* meetings further complicates the true understanding of the motives and relative power of constituencies involved in the policy process. Examining the FCC discourse helps reveal the key ideas and perspectives that frame the acceptable boundaries of policymaking itself. This dramaturgy is quite different from that found in the radio industry trade press, where discourse is oftentimes more frank than that found in the policy arena. Carefully examining this forum is useful for ferreting out the "truth" (or lack thereof) of the various constituents involved in the policy process.

There are several key actors in the digital radio dramaturgy. *The FCC* is both a stage and actor, responsible for setting and maintaining the groundrules of policy discourse. Surprisingly, the FCC engages in little public dialogue regarding digital radio, instead deferring to preferred constituents in an attempt to promulgate what it perceives to be the "best" policy, as biased by its own historical and institutional understanding of radio regulation. *The industry* encompasses all corporate proponents of radio's digital transition: this includes iBiquity Digital Corporation, the proprietor of the U.S. "HD Radio" technology; the nation's largest broadcast conglomerates—all of whom are investors in iBiquity and collectively control the majority of radio industry revenues; the National Association of Broadcasters, which has worked to facilitate a semblance of inevitability surrounding the adoption of digital radio; and public broadcasters, who were instrumental in advancing radio's digital transition by leveraging the institutional credibility they have with policymakers. Industry actors are the primary protagonists in this particular dramaturgy, as they direct the "plot" toward an outcome most advantageous to themselves. Furthermore, the policy arguments the industry advances are the ones against which the FCC judges all other perspectives.

Independent broadcasters are members of the broadcast community who are materially unaffiliated with those constituencies responsible for the development and promulgation of digital radio. They are the primary antagonists in this dramaturgy. Although independent broadcasters represent the majority of radio stations in the United States, they've strongly opposed the operative rationales employed by the industry to define the future of radio. Relatedly, *consulting engineers* are broadcast-certified scientists not tied to a single station or broadcast conglomerate. Initially, they worked to concretize the industry's operative arguments about digital radio, but later openly questioned the HD system's viability. *Electronics manufacturers,* which include the makers of broadcast transmission and reception equipment as well as end-users of such components, such as automobile manufacturers, are also important antagonists in this dramaturgy. One might think that electronics manufacturers would be closely aligned with the broadcast industry on the issue of digital radio, but instead they exhibited reticence to its adoption. Finally, *the public* also played an antagonistic role. Whether it be individual citizens with the knowledge and desire to participate in the policymaking process, or nonprofit groups who intervene in issues of communications policy on behalf of the public, they (much like independent broadcasters) advanced important arguments that were oppositional to stated industry objectives regarding digital radio.

The actors in the technological and policy development of digital radio in the United States came to conflict over several discursive threads. Dominant among them is the idea that *it is inevitable that radio become a digital medium.* Stemming from industry anxiety about convergence, as well as an institutional orientation within the FCC to embrace new technologies without a full understanding of their implications, cementing this notion early on was integral to how the U.S. digital radio transition would be conducted. Secondly, *digital radio need not be significantly superior to its analog equivalent*; why this must be so is directly related to the form of technology preferred by the broadcast industry, which simply does not have the capability to provide radio with meaningfully qualitative improvements in service. Furthermore, *sacrifices must be readily made to analog radio service in order for digital broadcasting to succeed.* This is the most contentious thread of discourse because it hints at the potentially disruptive nature of digital radio technology and makes the premature assumption that there is little value left in analog broadcasting.

Examining the dramaturgy of digital radio's development in the United States paints a very revealing picture of where power is located within the process of contemporary communications policymaking and guarantees that the voices of all the actors involved in the process are fairly heard in proper context. Furthermore, comparing words with deeds highlights the inherent weaknesses of radio's digital transition in the United States and illuminates just how far detached modern broadcasting really is from constructively navigating the convergence phenomenon.

ASSEMBLING THE STORY OF RADIO'S DIGITAL DILEMMA

This book utilizes several archival sources to tell the story of the U.S. digital radio transition, using the words of participants themselves to illustrate the development of technology and policy. This topic has received scant attention from scholars, so there is little direct academic foundation on which to start. That said, the transition process has been underway for more than 20 years now, so there is no shortage of material with which to work. This includes the entirety of the FCC's rulemaking record on terrestrial digital radio; two dockets in particular (RM-9395 and MM 99-325) constitute the entire archive of proposals, comments, and decisions promulgated by the FCC to facilitate the transition. Both are available online via the FCC's Electronic Comment Filing System (ECFS).[73] Some 1,500 unique filings are available in both dockets, ranging in length from a paragraph to 700+ pages apiece. These filings give a clear picture of the motivations and arguments proffered by various constituents in the policy dramaturgy and are extremely important for illuminating the operative rationales that have controlled (or conflicted with) policy outcomes.

In addition, the archives of industry trade publications provide important context to the policy dramaturgy of digital radio—of which two in particular are especially enlightening. *Radio World* is the preeminent industry newspaper for broadcast station owners, managers, and engineers.[74] Published biweekly, each 50-page issue provides an excellent forum in which to observe the discursive behavior of actors directly involved in the radio industry, whether they are protagonists or antagonists in a policy context. Although *Radio World* relies wholly on industry advertising for support and, thus, displays an editorial bias in favor of industry desires, the publication's change in tenor over time with regard to the potential of digital broadcasting can be clearly mapped from 1988 onward, which makes it a useful lens through which to observe how struggles over digital radio developed and festered outside the realm of the official policy discourse. Similarly, the trade publication for public radio, *Current,* was examined to glean information about noncommercial broadcasters' support for and concerns with digital radio.[75] A smaller publication (20–30 pages per issue), *Current* is also published biweekly, and its archives were also analyzed from the late 1980s through the present. Ultimately, the dialogue in *Radio World, Current,* and other publications serves to supplement the FCC record with more candid assessments of digital radio technology and its prognosis for success—highlighting the fundamental contradictions between what digital radio broadcast technology is purported to be and what it actually is. These contradictions lie at the heart of radio's digital dilemma, both from an operational and a policy standpoint.

Radio's Digital Dilemma comprises eight chapters in all. Chapter 1 describes the theoretical and methodological underpinnings of this work, as well the source material from which it is constructed. Chapter 2 charts the initial developmental trajectory of U.S. digital radio broadcasting. Covering a timeline between 1988 and 1998, it summarizes the extensive discussions that took place in the trade press about digital radio before the FCC became directly involved with the issue. During this phase, definable constituencies in the forthcoming policy process began to congeal, focused around three pivotal moments in the decade: the constitutive decision to pursue a home-grown digital radio technology; the imposition of the Telecommunications Act of 1996 and its effects on the radio industry; and the formal commencement of an FCC rulemaking on the subject. Surprisingly, this period illustrates deep-seated disagreement between various actors about the necessity for and viability of digital radio broadcasting itself. These early schisms, and their subsequent fallout, would significantly affect how the policy dramaturgy would evolve.

Chapter 3 provides a critical overview of the fundamental deficiencies of U.S. digital radio technology. This critique can be boiled down to three main categories. The first involves spectral integrity: the intermixing of analog and digital radio signals on the AM and FM dials not only strengthens the property rights of incumbent broadcasters, but has also demonstrated potential to degrade existing analog broadcast service. The second critique involves bandwidth capacity: although the designers of HD Radio proclaim it can be used to provide a panoply of new services, in reality it neither extends the medium's communicative effectiveness nor provides it with a useful vector into a convergent media environment. Finally, the wholly proprietary nature of the U.S. digital radio standard is explained. This arrangement raises the specter of a future where two licenses will be required by broadcasters: one from the FCC for the use of the public airwaves, and another from iBiquity Digital Corporation for the permission to transmit digitally. Not only does this represent a significant shift in the gatekeeper function of the FCC itself, but a closed system also stifles innovation and inflates adoption costs for both the radio broadcaster and listener.

The analysis of the digital radio policy process begins in earnest with Chapter 4. Proponents of HD Radio begin to deploy, refine, and buttress their key arguments for adoption of the technology, while those opposed conduct a thorough and critical analysis of its shortcomings. Despite a lack of substantive, impartial, and independent testing, the FCC approved the deployment of HD Radio in 2002. How this process played itself out speaks volumes about the relative disengagement of the FCC regarding the development and testing of digital radio, and the power of neoliberal ideology over the adoption of new communications technology and policy.

Chapter 5 chronicles the initial rollout of digital radio in the United States. During this period, serious technical faults are found with HD Radio when deployed in the real world, which generates a flurry of intra-industry discussion about whether or not digitalization is actually good for the medium itself. Independent broadcasters, members of the public, and consulting engineers all bring forth compelling evidence of digital radio's significant shortcomings and articulate alternative futures for the medium. Some radio stations that were early-adopters abandon the technology, and HD proponents work furiously to engineer fixes while petitioning the FCC not to intervene. In fact, regulators approve further rules governing digital radio deployment that *loosen* restrictions on its proliferation and exacerbate its real-world consequences.

Chapter 6 explores the radio industry's desperate attempt to reengineer the regulation of digital radio in order to fix one of its fundamental detriments—the fragility of HD Radio signals themselves. After nearly six years on the air, HD proponents petition the FCC to allow FM stations to increase the power of their digital signals by tenfold; this inflames consternation among many independent broadcasters and the listening public, who worry that such a power increase will interfere with analog stations, many of which opted out of HD Radio. Much like the service's initial rollout, broadcast conglomerates and public broadcasters hammer out a "compromise" proposal to justify the power increase, which the FCC unquestioningly adopts—turning decades of spectrum-integrity policy on its head in the process. This struggle-within-a-struggle epitomizes neoliberalism's triumph in communications policymaking more generally.

Chapter 7 reviews the current state of U.S. digital radio. The number of HD-capable stations on the air has stagnated and may be in decline. Transmitter manufacturers lament the lack of demand; consumer electronics companies, reticent about HD technology from the outset, refuse to mass-produce compatible receivers. The detrimental attributes of HD Radio signals cause consumer confusion and dismay; listener interest is also in a state of malaise. As a result, the term "digital radio" itself has evolved to encompass other services such as satellite radio and streaming audio delivered via wireless broadband. In response, HD proponents have embarked on a variety of developmental and promotional campaigns to try and provide the technology with some semblance of viability, but the scale and scope of these campaigns is questionable and some may even be working at cross-purposes.

The concluding chapter (Chapter 8) highlights the weakness of substance on which digital radio policies are based, offers lessons from the story of radio's digital transition that may be useful in the broader world of convergent media policy, and contemplates the future of radio broadcasting in the context of the confusion that exists over its many paths to digitalization. Ultimately, radio's digital dilemma invites a radical rethinking of radio's identity and relative importance in our increasingly convergent media environment. For those who still find value in radio broadcasting, it is not the concept of the medium that needs reclamation, but, rather, its purpose.

NOTES

1. Henry Jenkins, *Convergence Culture: Where Old and New Media Collide* (New York: New York University Press, 2006), 243.
2. Dan Schiller, *How to Think About Information* (Urbana: University of Illinois Press, 2007), 103.
3. Andrew Calabrese, "Stealth Regulation: Moral Meltdown and Political Radicalism at the Federal Communications Commission," *New Media & Society* 6, no. 1 (February 2004): 112.
4. Lisa Gitelman, *Always Already New: Media, History, and the Data of Culture* (Cambridge, MA: MIT Press, 2006), 93.
5. See Jeff Chester, *Digital Destiny: New Media and the Future of Democracy* (New York: The New Press, 2007), xvii–xviii; Schiller, 139; and Edward S. Herman and Robert W. McChesney, *The Global Media: The New Missionaries of Global Capitalism* (London: Cassell, 1997), 134, 198.
6. David Sedman, "Radio Transmission," in Peter B. Seel and August E. Grant, eds, *Broadcast Technology Update: Production and Transmission* (Boston: Focal Press, 1997), 157.
7. Ibid., 158–159.
8. For an overview of neoliberalism, its origins, and its entrenchment in public policymaking more generally, see David Harvey, *A Brief History of Neoliberalism* (New York: Oxford University Press, 2007).
9. See Marko Ala-Fossi and Alan G. Stavitsky, "Understanding IBOC: Digital Technology of Analog Economics," *Journal of Radio Studies* 10, no. 1 (2003): 63–79; Andrea Baker, *Virtual Radio Ga Ga, Youths, and Net-Radio: Exploring Subcultural Models of Audiences* (New York: Hampton Press, 2012), 118–119; R.L. Hilliard and M. C. Keith, *The Quieted Voice: The Rise and Demise of Localism in American Radio* (Carbondale: Southern Illinois University Press, 2005), xxii; and Stephen Lax, Marko Ala-Fossi, Per Jauert, and Helen Shaw, "DAB: The Future of Radio? The Development of Digital Radio in Four European Countries," *Media, Culture, & Society* 30, no. 2 (2008): 152.
10. See Ala-Fossi and Stavitsky, 74; and John Anderson, "Digital Radio in the United States: Privatisation of the Public Airwaves?" *Southern Review: Communication, Politics and Culture* 39, no. 2 (2006): 5.
11. See David S. Allen, *Democracy, Inc.: The Press and the Law in the Corporate Rationalization of the Public Sphere* (Urbana: University of Illinois Press, 2005), 2; Nicholas Garnham, *Capitalism and Communication: Global Culture and the Economics of Information*, edited by Fred Inglis (London: Sage,

1990), 111; Ernest A. Hakanen, "On Autopilot Inside the Beltway: Organizational Failure, the Doctrine of Localism, and the Case of Digital Audio Broadcasting," *Telematics and Informatics* 12, no. 1 (1995): 11, 15; Robert W. McChesney, *The Political Economy of Media: Enduring Issues, Emerging Dilemmas* (New York: Monthly Review Press, 2008), 157, 247, 345, 371; Robert W. McChesney, *Rich Media, Poor Democracy: Communication Politics in Dubious Times* (New York: The New Press, 2000), 124–125; Herman and McChesney, 7; Schiller, 55; Thomas Streeter, *Selling the Air: A Critique of the Policy of Commercial Broadcasting in the United States* (Chicago; University of Chicago Press, 1996), 20; and Frank Webster, *Theories of the Information Society, Third Edition* (London: Routledge, 2006), 270–271.

12. See Francois Fortier, *Virtuality Check: Power Relations and Alternative Strategies in the Information Society* (London: Verso, 2001), 103; Hakanen, 15; and Armand Mattelart, *Networking the World: 1794–2000*, translated by Liz Carey-Libbrecht and James A. Cohen, (Minneapolis: University of Minnesota Press, 2000), 104–105.

13. Ala-Fossi and Stavitsky, 64–65.

14. Paul Starr, *The Creation of the Media: Political Origins of Modern Communications* (New York: Basic Books, 2004), 4.

15. Ibid., 5.

16. See Mark Lloyd, *Prologue to a Farce: Communication and Democracy in America* (Urbana: University of Illinois Press, 2006), 119; Robert W. McChesney, *The Problem of the Media: U.S. Communication Politics in the Twenty-First Century* (New York: Monthly Review Press, 2004), 47, 226; Robert W. McChesney, *Telecommunications, Mass Media, and Democracy: The Battle for Control of U.S. Broadcasting, 1928–1935* (New York: Oxford University Press, 1993), 10, 264–265; Susan Smulyan, *Selling Radio: The Commercialization of American Broadcasting, 1920–1934* (Washington, D.C.: Smithsonian Institution Press, 1994), 6, 162, 166; and Streeter, 37.

17. See Patricia Aufderheide, *Communications Policy and the Public Interest: The Telecommunications Act of 1996* (New York: Guilford Press, 1999), 8; McChesney, *Telecommunications, Mass Media, and Democracy*, 92–187, 196–225, 261; Smulyan, 1; and Streeter, 275.

18. McChesney, *The Political Economy of Media*, 179.

19. Starr, 372.

20. Streeter, 37.

21. Ibid., 39.

22. Ibid., 128.

23. See Robert Britt Horwitz, *The Irony of Regulatory Reform: The Deregulation of American Telecommunications* (New York: Oxford University Press, 1989), 36, 44; Sven B. Lundstedt and Michael W. Spicer, "Latent Policy and the Federal Communications Commission," in Sven B. Lundstedt, ed., *Telecommunications, Values, and the Public Interest* (Norwood, NJ: Ablex Publishing Corporation, 1990), 290–292, and McChesney, *Telecommunications, Mass Media, and Democracy*, 28–29.

24. Susan J. Douglas, *Inventing American Broadcasting, 1899–1922* (Baltimore: Johns Hopkins University Press, 1987), 236–237.

25. C. Twight, "What Congressmen Knew and When They Knew It: Further Evidence on the Origins of U.S. Broadcasting Regulation," *Public Choice* 95, no. 3/4 (1998): 269–271.

26. See McChesney, *Telecommunications, Mass Media, and Democracy*, 29, 255; and Starr, 349, 351.

27. Hugh Richard Slotten, "Rainbow in the Sky: FM Radio, Technical Superiority, and Regulatory Decision-Making," *Technology & Culture* 37, no. 4 (October 1996): 709.
28. See Patricia Aufderheide, *Communications Policy and the Public Interest: The Telecommunications Act of 1996,* 61, 105–106; and Patricia Aufderheide, "Shifting Policy Paradigms and the Public Interest in the U.S. Telecommunications Act of 1996," *The Communication Review* 2, no. 2 (1997): 272, 274–275.
29. See McChesney, "Forward," in Hilliard and Keith, x; and McChesney, *The Problem of the Media,* 45.
30. Garnham, 127.
31. Nicholas Garnham, "Information Society Theory As Ideology: A Critique," *Studies in Communication Sciences* 1 (2001): 164.
32. Starr, 352, 362–373.
33. Mark J. Braun, *AM Stereo and the FCC: Case Study of a Marketplace Shibboleth* (Norwood, NJ: Ablex, 1994), 167.
34. Walter B. Emery, *Broadcasting and Government: Responsibilities and Regulations* (East Lansing: Michigan State University Press, 1971), 292.
35. See Braun, 141; Hernan Galperin, *New Television, Old Politics: The Transition to Digital TV in the United States and Britain* (Cambridge, UK: Cambridge University Press, 2004), 7, 70, 244; Horwitz, 38–39, 48, 88, 155; Erwin G. Krasnow and Lawrence D. Longley, *The Politics of Broadcast Regulation* (New York: St. Martin's Press, 1973), 80–81; Vincent Mosco, *Broadcasting in the United States: Innovative Challenge and Organizational Control* (Norwood, NJ: Ablex Publishing Corporation, 1979), 5, 46, 61, 126–127; and Philip M. Napoli, *Foundations of Communications Policy: Principles and Process in the Regulation of Electronic Media* (Cresskill, NJ: Hampton Press, 2001), 75–77, 215–216, 273.
36. See Chester, 46–64; John Dunbar, "Who Is Watching the Watchdog?" in Robert W. McChesney, Russell Newman, and Ben Scott, eds, *The Future of Media: Resistance and Reform in the 21st Century* (New York: Seven Stories Press, 2005), 131; Lundstedt and Spicer, 295; and Hugh Richard Slotten, *Radio and Television Regulation: Broadcast Technology in the United States, 1920–1960* (Baltimore: The Johns Hopkins University Press, 2000), 62–63.
37. See Braun, 77–78; Emery, 301; and Napoli, 88–90.
38. Dunbar, 139.
39. Lundstedt and Spicer, 292–293.
40. Philip M. Napoli and Joe Karaganis, *Toward a Federal Data Agenda for Communications Policymaking* (New York: Social Science Research Council, 2007), 6.
41. Napoli, 268–269.
42. Napoli and Karaganis, 7.
43. Napoli, 123.
44. Daniel Bell, *The Coming of Post-Industrial Society: A Venture in Social Forecasting* (New York: Basic Books, 1973), 364–365.
45. Sandra Braman, *Change of State: Information, Policy, and Power* (Cambridge, MA: The MIT Press, 2006), 319.
46. Napoli, 266.
47. Ibid., 254.
48. Transcript of remarks delivered by FCC Commissioner Jonathan Adelstein, Media Policy Research Pre-Conference to the National Conference for Media Reform, Memphis, TN, January 11, 2007, http://katrinaresearchhub.ssrc. org/media-hub/news/transcript-of-fcc-commissioner-adelsteins-remarks-at-media-policy-research-pre-conference/.

49. Eric Klinenberg, *Fighting For Air: The Battle to Control America's Media* (New York: Metropolitan Books, 2007), 277–278.

50. Matthew Lasar, "FCC Insider: This Place Is Hell; Silent Protest Planned," *Ars Technica*, March 16, 2008, http://arstechnica.com/tech-policy/2008/03/fcc-insider-this-place-is-hell-silent-protest-planned/.

51. See Hilliard and Keith, 59–63; Klinenberg, 203–207; and Krasnow and Longley, 36–37.

52. Napoli and Karaganis, 15.

53. Napoli, 73, 233.

54. See JoAnne Holman and Michael A. McGregor, "'Thank You for Taking the Time to Read This:' Public Participation via New Communication Technologies at the FCC," *Journalism and Communication Monographs* 2, no. 4 (December 2000): 162, 164, 182; and Michael A. McGregor, "When the 'Public Interest' Is Not What Interests the Public," *Communication Law and Policy* 11, no. 2 (Spring 2006): 210.

55. See McGregor, 209, 223–224; and Holman and McGregor, 185, 187.

56. See Napoli, 230; and Robert W. McChesney, *Digital Disconnect: How Capitalism Is Turning the Internet Against Democracy* (New York: The New Press, 2013), 93–95.

57. Krasnow and Longley, 26.

58. Braman, 325.

59. Napoli, 226.

60. McChesney, *The Political Economy of Media*, 235.

61. Ibid., 353.

62. Ibid., 350.

63. See Erving Goffman, *The Presentation of Self in Everyday Life* (New York: Doubleday, 1958); and Erving Goffman, *Frame Analysis* (New York: Harper Books, 1974).

64. See Peter M. Hall, "Asymmetry, Information Control, and Information Technology," in David R. Maines and Carl J. Couch, eds, *Communication and Social Structure* (Springfield, IL: Charles C. Thomas, 1988), 355; Jan Ekecrantz, "Collective Textual Action: Discourse, Representation, Dramaturgy and Public Interaction in the Media Sphere," *Nordicom Review* 18, no. 2 (Nov. 1997): 21; and Dan Kärreman, "The Scripted Organization: Dramaturgy from Burke to Baudrillard," in Robert Westwood and Stephen Linstead, eds, *The Language of Organization* (London: Sage, 2001): 108.

65. Ekecrantz, 38.

66. Peter K. Manning, "The Truthfulness of Organizational Communication," in David R. Maines and Carl J. Couch, eds, *Communication and Social Structure* (Springfield, IL: Charles C. Thomas, 1988), 101.

67. A more substantive treatment of the history and practicality of dramaturgical analysis can be found in John Anderson, *Radio's Digital Dilemma: Broadcasting in the 21st Century* (Doctoral dissertation, University of Illinois at Urbana-Champaign, 2011), 14–20.

68. Ibid., 106–107.

69. Ibid., 109.

70. Hall, 350–351.

71. Streeter, 114.

72. Ibid., 128, 148.

73. To examine the docket of MM 99–325, visit http://apps.fcc.gov/ecfs/comment_search/; click the link to remove the date restriction, and type "99–325" or "RM-9395" (without quotes) into the "Proceeding" field.

74. *Radio World* also publishes a significant amount of content online at http://www.rwonline.com/.

75. *Current* also publishes online; see http://www.current.org/.

2 The Developmental Trajectory of U.S. Digital Radio

The constitutive decisions regarding radio's digital transition were made during the 1990s. At first, U.S. broadcasters tentatively agreed to adopt the same digital radio technology as the rest of the industrialized world; however, reticence among regulators to allocate new spectrum necessary for the service effectively forced broadcasters to invent their own system that could be deployed on the existing AM and FM bands. As research into "in-band" digital broadcasting progressed, the radio industry supported developers based not on the merits of their work, but on who financed their work. Two major proponents of in-band technology would emerge, but only one would be selected as the U.S. standard. Not surprisingly, the developer with the strongest industry ties ended up dominating the process. The FCC was wholly uninvolved in digital radio development and did nothing during the decade to constructively advance the issue. Had regulators been actively engaged in the early stages of digital radio research, the technology ultimately chosen might have represented a significant advancement in broadcasting more generally; instead, the FCC allowed incumbent broadcasters to construct a system that served their own economic interests first, and effectively ignored all other metrics by which to judge its actual viability.

Early prognostications about what digital radio would be had no basis in empiricism: by the time the FCC was presented with a technological framework, its proprietors had only proven that the simultaneous transmission of both analog and digital radio signals was possible. During the early stages of research, broadcasters identified satellite radio as the primary competitive threat to terrestrial broadcasting that would eat away at their listening audiences and, in turn, cut into their revenue streams. They were fearful of the fact that satellite radio would be digital from its inception, while terrestrial radio remained analog. Thus, broadcasters' desire for digital radio was based on the concept of digitalization itself, not on any tangible benefits that digitalization might provide.

This created an early schism between two constituencies necessary for the successful adoption of any digital radio technology—broadcasters and receiver manufacturers. The latter did not support the concept of in-band digital broadcasting; this solution would require electronics manufacturers to produce entirely new radio receivers for the domestic market. They tried repeatedly throughout the decade to advance the notion of a global digital radio standard, while broadcasters favored a technology that, while unproven, would give them firm control over the pace and principles of any digital transition.

Since the FCC would not formally engage in the digital radio debate until 1998, the work of the decade—and its rationalization—would be articulated primarily within the trade press. In the case of *Radio World*, coverage would focus on the companies working to develop in-band technology, and the sentiments among commercial broadcasters about which developer seemed to offer the most profitable and least disruptive method by which radio stations might go digital. *Radio World* also extensively covered the period of trial-and-error that occurred during the early days of in-band development, but always from the perspective of unarticulated "progress," defined by the somewhat questionable dictum that "digital equals better." Problems that arose during the development process were characterized as growing pains, as opposed to possible red flags that an in-band solution could not realistically provide broadcasters with a robust analog-to-digital transition path. *Current*, the trade publication of public broadcasters, cast the role of public radio as an innovator within the digital radio space; initially, claims about the technology's functionality were more grandiose in *Current* then they were in *Radio World*. As it became clear that a commercial broadcaster-endorsed technology would prevail, public broadcasters focused primarily on preventing their marginalization in any digital radio transition.

PREMATURE CONSENSUS IN THE FACE OF SATELLITE RADIO

On August 1, 1986, public radio broadcaster WGBH-FM in Boston successfully simulcast its programming over a digital sideband adjacent WGBH-TV's analog signal: the first documented instance of a U.S. radio station making a tentative foray into the world of digital audio broadcasting.[1] The test had no significance beyond proving the theory that digital audio could be transmitted as its own unique data stream. Other than this single test, the U.S. radio industry did not actively pursue digital radio research until the FCC forced the issue in 1990 as part of a wide-ranging proceeding on digital audio services.[2] The rulemaking centered on the creation of a satellite-delivered digital radio service, but left open the possibility for setting fundamental ground rules regarding the development of terrestrial digital broadcasting.

In reaction to the FCC's rulemaking, the industry convened a private working group of commercial and public broadcasters to contemplate radio's digital future. Mike Starling, one of National Public Radio's top engineering experts, co-chaired this initiative. The group sponsored a seminar for U.S. and European broadcasters to discuss terrestrial digital radio;[3]

by the end of 1990, Starling claimed that "from all indications, [the adoption of a digital radio standard] is on the fast track at the FCC."[4] John Wilner, a staff writer for *Current*, explained the importance of NPR taking an early leadership role in digital radio development: "In an effort to avoid the mistakes of the 1920s, when what was then called 'educational' radio was tacked onto the nascent commercial broadcasting system as an afterthought, NPR and other public radio organizations are eager to secure a position in the digital movement." Said Starling, "Our support for [digital broadcasting] has really been based on positioning public radio so that we can take advantage of it and participate in it. The only way to compete with those people is to become one of them."[5]

Initially, the U.S. broadcast industry seemed prepared to adopt the apparent global digital radio standard, then known as Eureka 147 DAB, and the National Association of Broadcasters (NAB) formally passed a resolution pursuing domestic study and development of Eureka at its annual convention in 1991.[6] The Eureka system was conceived in 1981 by a German-based consortium, with research and development assistance from the country's corporate sector and several European public service broadcast agencies.[7] Notably, Eureka 147 worked on swaths of spectrum not allocated to radio broadcasting at the time. Initial testing of the Eureka system began in 1985 and was formalized in 1986; its first field demonstration took place in Geneva, Switzerland, in 1988.[8]

Unlike traditional, terrestrial analog broadcasting, the Eureka 147 system is designed around the use of "multiplexes," or single transmitters that occupy a range of spectrum in a given geographic area. Individual program providers are offered slices, or "channels," of spectrum available on a multiplex. These channels provide each station with the capability of delivering a high-quality audio stream, or the option of carving up its channel into multiple program streams, not all of which may be audio-based.[9]

System specifications for Eureka 147 were finalized in 1994, and the International Telecommunication Union (ITU) subsequently recommended it as a global digital radio standard.[10] The British Broadcasting Corporation (BBC) and Swedish Broadcasting Corporation launched the first full-time Digital Audio Broadcasting (DAB) services in September 1995;[11] other countries within Europe and beyond began active pilot projects using the Eureka platform.[12] After lining up the commitment of several broadcasters and receiver manufacturers,[13] Canada formally adopted the Eureka system;[14] its transition was driven by the potential of "value added services," such as datacasting, that digital radio might bring to the broadcast marketplace.[15] By 1997, the Eureka system was considered "a de facto standard in Europe, Canada, Mexico, and many other countries,"[16] and the president of Eureka's European promotional efforts, David Witherow, confidently predicted that the system was "on its way to becoming—if it is not already—a world standard for digital radio."[17]

U.S. problems with adoption of the Eureka system began shortly after the industry commenced a nationwide spectrum inventory for digital radio. At the time, the frequencies on which the Eureka system could be deployed were in geographically sporadic but systematic use for military applications, such as flight- and missile-test telemetry.[18] Within two months of the NAB's resolution endorsing Eureka 147, the Pentagon formally objected to the notion of appropriating the necessary spectrum for digital broadcasting.[19] A short but intense flurry of lobbying ensued: the FCC suggested three other patches of spectrum for terrestrial digital radio using Eureka technology, but all were shot down by the National Telecommunications and Information Administration (NTIA)—the regulatory body charged with overseeing the use of spectrum for government and military purposes.[20] An attempt to legislatively force a spectrum set-aside, mostly carried forth by public broadcasters, also failed in Congress.[21]

By January 1992, the radio industry realized it did not have the political muscle to wrest new spectrum from the military-industrial complex, and the FCC seemed unwilling to force the issue on its behalf. This resulted in the effective abandonment of *any* form of alternate-band digital radio in the United States.[22] Seeing no immediate alternative to Eureka on the horizon, the FCC terminated its terrestrial digital radio proceeding.[23] Its rulemaking now focused solely on satellite, from which the SiriusXM radio service was born.[24] This provoked much consternation within the radio industry,[25] which viewed the proposed launch of satellite radio later in the decade as a make-or-break deadline by which terrestrial broadcasters also needed to go digital.[26] By detaching digital terrestrial radio from its larger rulemaking, the FCC effectively ceded control of its development to broadcasters exclusively.

The looming rise of satellite radio worried broadcasters in two identifiable ways. First, it promised increased programming choice, which the radio industry feared would entice listeners away from their stations. Fewer listeners would translate into lower station ratings and a concomitant depression in advertising rates, thereby fiscally weakening the entire broadcast industry. Secondly, satellite radio would be a fully digital service, and the working assumption at the time was that these broadcasts would provide a higher audio quality than did traditional analog AM and FM stations. Although both of these assumptions had no empirical evidence to support them at the time, and despite a fundamental access distinction between terrestrial and satellite radio (the latter requires a monthly subscription, while the former is free to anyone with a receiver), satellite radio was considered threatening enough to compel the broadcast industry to begin its own digital radio research and development initiatives. Since Eureka 147 was the only system in existence at the time, the industry would have to create its own from whole cloth.

"COMING TO GRIPS" WITH IN-BAND DIGITAL RADIO

Even though the NAB had initially endorsed Eureka 147, the system did not have the unanimous support of its membership. Not only would the process of multiplexing harmonize the transmission power (and resultant coverage area) of all competing stations in a market, but broadcasters also feared "a marketer's nightmare" in trying to educate the listening public about any alternate-band transition.[27] The fear of new competition in a multiplexed environment probably played as much of a role in quashing the notion of alternate-band digital radio in the United States as did the military's opposition.[28] In fact, there were some corporations, such as Westinghouse, which had interests in both broadcasting and the military: Westinghouse had invested millions of dollars in avionics infrastructure that used spectrum now proposed for digital broadcasting.[29] There is also evidence that the NAB made the decision to endorse Eureka 147 out of self-interest: under the tentative agreement reached (but never ratified) by the NAB and the Eureka 147 consortium, the NAB would have "developed plans for the association's profit-making arm to license the technology to American stations for a fee." This would have been a radically new role for a trade association that ostensibly worked to sustain the overall economic health of its membership.[30]

The record also shows that the U.S. radio industry had not put all of its eggs in the Eureka basket. Westinghouse founded an alternative digital radio development effort, code-named "Project Acorn," in 1989. Started as a relatively informal science experiment, over the next two years the Gannett Corporation and CBS Radio would officially join the venture, while public radio executives and engineers also signed on. Project Acorn was an attempt to develop an "in-band" system—one where analog and digital signals would coexist on the traditional AM and FM broadcast bands. This would solve the problem of needing to move all stations to new spectrum; it would also maintain the variable coverage of stations under existing FCC allocation rules and, most of all, protect incumbents' investments in the existing radio broadcast marketplace.[31] Technically speaking, in-band digital broadcasting was previously considered impossible without harming existing analog radio services. The conundrum boils down to one of the most fundamental rules of radio-frequency physics: if you add new energy into an area of spectrum already occupied, interference is bound to result. Nevertheless, in-band digital radio would be officially endorsed as public radio's preferred digital transition plan in a Corporation for Public Broadcasting study released in 1993, and it expected FCC approval of the Project Acorn concept "within a year or so."[32] Public radio engineers suggested radio's digital transition would take, at minimum, 10 to 15 years, probably beginning with "simulcasts of our old medium on our new medium."[33] The NAB similarly expected any digital radio transition to take "a long time."[34]

At the same convention where the NAB debuted Eureka 147 with much fanfare, a muted display of Project Acorn's concept was also demonstrated.[35] The first display unit was actually a mockup, which could only illustrate what the effects might be of placing digital data underneath and around a typical FM broadcast signal. By 1992, Westinghouse, Gannett, CBS, and other interested broadcasters founded a separate company, USA Digital Radio (USADR), to consolidate their research and development efforts toward the creation of a workable in-band system. Housed as a subsidiary of Westinghouse and initially based in Chicago,[36] USADR initially "had no more than two or three employees," with the rest of its staff on loan from investor-companies.[37] It farmed out most of its early research to consultants and contract engineering laboratories.[38] The FCC took no active role in these early efforts.

That same year, the Corporation for Public Broadcasting (CPB) announced it would spend $350,000 on research into digital broadcasting. The CPB was optimistic about the ways in which in-band digital radio would allow broadcasters entre into a convergent digital media environment.[39] NPR's point-person on all things digital, Mike Starling, acknowledged that in-band DAB might bring the opportunity to introduce some new services to the radio environment, but these would be minor relative to broadcasting's primary role of audio content distribution.[40] More important to Starling was that public broadcasters had secured a seat at the table during the start of in-band development: "Digital radio is still at the talking stage in this country . . . We are extremely fortunate to have a more formalized presence in telecommunications policymaking than our predecessors . . . We have a blank sheet of paper for the design of the role of public radio in the American public's future."[41] How this would be actualized was, at the time, technically indescribable. The CPB itself suggested that prognostications about digital radio's potential were premature, "because technology only defines what *can* be—the consumer's wants and needs, the interaction of competing products in a marketplace, economics, finance and regulation collectively define what *will* be."[42] This did not stop some commentators from predicting a fundamental shift in radio's identity. Some suggested "tailored bit-streaming" would become a major application, and display screens and printer ports would subsequently be installed in receivers, morphing them into multimedia devices.[43]

At 4 a.m. on August 29, 1992, public radio station WILL-FM in Urbana, Illinois became the first in the United States to successfully conduct an experimental hybrid analog/digital broadcast.[44] The prototype hardware was developed at an unnamed laboratory on the campus of the University of Illinois at Urbana-Champaign, to which a part of USADR's research had been contracted. When WILL-AM/FM/TV General Manager Donald Mulally inadvertently tumbled to the project, he "volunteered the FM station as a guinea pig." The first hybrid FM broadcast transmitted the digital portion of the signal at 1/1000th the power of WILL's analog signal.[45]

Engineers were so impressed with the proof-of-concept that the prototype was packed up and displayed at the National Association of Broadcasters' annual radio convention the following month. Unfortunately, only "one hand-built prototype receiver" could decode this first foray into in-band digital broadcasting, and nothing substantive was reported about its audio fidelity or signal robustness.[46] Meanwhile, in Cincinnati, Ohio, USADR began testing hybrid AM broadcasts on an experimental station.

Following these crude demonstrations, other companies launched their own research and development projects on in-band digital radio. Among them were AT&T and its then-subsidiary, Lucent Technologies.[47] Lucent began development of an in-band system as a means by which to spin off existing research on digital audio technology into markets that could be readily monetized.[48] While both USADR and Lucent had pounced on the idea of a hybrid analog/digital FM broadcast system, USADR was the only company initially working on an AM broadcast scheme.[49]

Although competition in the innovative phase of in-band DAB seemed like a positive development, the introduction of a second developer actually served to highlight the technical challenges of any in-band system and pushed constituents within the broadcast industry toward taking sides in the development process. Unlike USADR, which had the direct financial support of several major broadcasters and regularly tapped their insights on radio's digital future as part of its research, broadcasters saw Lucent as a disruptive "outsider" whose ideas about digital radio might not mesh with the industry that would ultimately deploy it. If a major goal of the radio industry was to have control over its digital future, it would be more likely to succeed under the auspices of a company it effectively owned than one in which it had no substantive relationship.

By 1994, the National Radio Systems Committee (NRSC) was engaged to oversee the testing of competing digital radio technologies, and established a Digital Audio Broadcasting Subcommittee to facilitate the process. The NRSC is a private consortium of representatives from the broadcasting, broadcast equipment, and consumer electronics industries. Typically, it is a place where industry works out consensus on new broadcast technology standards, and then forwards its endorsement to the FCC for formal evaluation and approval.[50] Discourse between NRSC members is private, with meetings closed to the public and press. With regard to the NRSC's DAB Subcommittee, the vast majority of its members were broadcasters who, incidentally, worked for companies that invested in USADR, and the subcommittee was chaired by a commercial broadcast engineer. The FCC did not engage in the NRSC's deliberations, effectively ceding any input or authority over the technology's development and testing processes, and signaling tacit approval that regulators would embrace any outcome. It took two years for the NRSC to develop in-band field-testing and system-evaluation guidelines.[51]

Notably, the first round of NRSC testing did *not* subject the competing technologies to a comparative analysis. Instead, the DAB Subcommittee simply set out to test their inherent viability. Initial criteria for an "adequate" digital radio system was defined as being robust to interference and other propagation characteristics (especially with regard to FM); it should offer "CD-quality [fidelity] standards"; it should not interfere with existing analog broadcast stations; and it should offer digital coverage equivalent to a station's analog service area. Questions about the deployment of ancillary services, such as datacasting, were not factored into consideration.

Early reports from the NRSC on laboratory testing of the proposed in-band systems "posted poor performers in every category."[52] One of the largest problems was the susceptibility of digital radio signals to interfere with their "host" analog stations. Bench-tests using a "representative" sample of radio receivers revealed this flaw in every system.[53] Given the disappointing results of the laboratory tests, preparations for the field-testing of in-band digital radio slipped into 1995.[54]

Despite the lack of a workable technology, USADR—with help from the NAB and trade press—began to paint a picture of inevitability about the adoption of in-band digital radio in the United States. At the Los Angeles World Media Expo in September 1994, USADR created a "static, in-booth demonstration" of its digital radio system, which included a controlled exhibition of hybrid analog/digital AM and FM broadcasts.[55] USADR representatives told Expo attendees that the company was already in discussion with equipment manufacturers to mass-produce transmission and reception components.[56] NAB DAB Task Force Chairman Alan Box declared USADR's system ready for deployment; the AM system, he boasted, "sounds as good as our current FM analog signals," while FM digital radio sounded as good as a CD. Other in-band proponents, such as Lucent, also had models of their equipment on display, and promised their own "static" demonstrations at forthcoming broadcast conventions.[57] None of these claims rested on any solid technical foundation. Early NRSC tests had conclusively proven that the developmental state of in-band digital radio was quite immature, and the technology did not yet meet the criteria necessary for viability, much less adoption.

In a *Radio World* commentary published in February 1995, NAB executive vice president John Abel attempted to articulate the potential of radio's digital future. In doing so, he cast digital broadcasting not as simply complimentary to existing analog radio services, but as a revolutionary mechanism by which broadcasters could expand the range of information they provide:

> The industry has yet to comprehend or understand what digital broadcasting means. To a large extent, radio broadcasters continue to think of "digital broadcasting" as a higher quality sound . . . But DAB is only one application of digital broadcasting: or to say it another way, DAB is simply an extension of the concept of digital broadcasting. Digital broadcasting *does not necessarily mean higher quality; digital broadcasting means the flexibility to achieve multiple purposes for the broadcast signal and certainly one of those applications is sound* [emphasis added].[58]

Predicting that "tomorrow's digital receivers will be more like today's computers," Abel insinuated that the reception of radio would merge into other devices, such as household appliances and mobile telephony, which would open up many new possibilities for broadcasting. He believed broadcasters could use the technology to create new "bit streams . . . allocated to e-mail paging, PDAs, signaling devices like utility load management, data transmissions, fax transmissions, differential global positioning system (D-GPS)" and the like. "Theoretically, all of these additional transmissions could be accomplished while still providing a real-time broadcast as is done now."[59]

Abel's commentary suggested that broadcasters hoped digital radio would not only give them qualitative parity to future satellite radio services, but would also open up new revenue streams related to digital media distribution more generally. Such hypothetical benefits glossed over the known technical deficiencies of in-band digital radio development, and urged the formation of consensus among broadcasters that a digital future was not only inevitable, but could be quite lucrative. Although there was no scientific basis for these claims, neither USADR nor Lucent cautioned against such wishful thinking. For its part, the FCC remained mum on the issue.

During the annual NAB convention in 1995, Las Vegas public radio station KUNV-FM was outfitted to broadcast a hybrid analog/digital signal,[60] while a hybrid AM demonstration-station broadcast live from the convention floor. USADR gave half-hour bus tours around carefully planned routes within the city to demonstrate the viability of its system.[61] Although proponents admitted that the AM side of the technology needed more work before market-size tests could begin, USADR strongly asserted that AM broadcasters were guaranteed a meaningful place in any digital radio transition.[62] In fact, the AM demonstration unit on the NAB convention floor had been assembled only two weeks before the show.[63] Unbeknown to attendees, USADR was collecting field data from these demonstrations, which it would later submit to the NRSC for evaluation.[64]

At the convention, Milford K. Smith, vice president of engineering for Greater Media, Inc., a USADR investor, led a roundtable on the promises of in-band digital radio.[65] It was not without fireworks: European engineers could not understand why the United States appeared to be bucking the global trend toward Eureka 147, and worried about the potential for interference that would be created by intermixing analog and digital radio signals.[66] NRSC Chairman Charles Morgan, then a vice president of engineering with Susquehanna Radio Corporation (another USADR investor), candidly responded, "We are basically studying to determine whether an in-band on-channel system will be substantially better than what we have today."[67] This was a far cry from the potential of digital radio envisioned by Eureka 147 supporters, and a much lower standard than the NRSC's own evaluative criteria for a successful system.[68]

Elsewhere at the NAB convention, USADR announced tentative agreements with a transmitter and semiconductor manufacturer to produce hardware for its embryonic technology.[69] All FCC commissioners, save Chairman

Reed Hundt, took the company's bus ride, though Hundt did inquire as to just how much "new spectrum" the system would require.[70] The response to this critical question could not be found on the bus, though Tony Masiell of CBS Radio—one of USADR's corporate founders—did admit that radio stations would need to expand their spectral footprints in order to accommodate the broadcast of both analog and digital signals. The implications of this on the traditional spectral allocation regime of the U.S. radio broadcast system went unquestioned at the time.[71] European attendees of the NAB convention good-naturedly challenged their U.S. counterparts to conduct a head-to-head test of in-band digital radio and the Eureka 147 system, though all agreed that any in-band technology required further refinement to make for a meaningful comparison.[72]

Some convention attendees who monitored USADR's demonstration stations on their own noted "splatter," which affected both AM and FM frequencies near the experimental outlets. Remarked one observer, "channel splatter of the magnitude that was present at the show would cause chaos in those markets where there are a lot of stations. I hope USA Digital Radio is also aware of these problems and has some plans to address them."[73] Robert C. Tariso, chief engineer of WLTW-FM in New York, was more emphatic:

> We will . . . have to come to grips with whether analog AM and FM can really coexist with this additional RF signal . . . no broadcaster in his or her right mind will destroy his or her existing business in the hope that digital will catch on . . . Let's try and understand more fully what we have . . . before we wipe the slate clean and start over. If we do not, perhaps the same laws of physics will plague a digital system as well. Trading one set of problems for another may not prove to be beneficial.[74]

These early rumblings of unease were brushed aside in the trade press. Several *Radio World* reports and editorials suggested that the continuation of research into other digital radio systems, such as Eureka 147, simply delayed the eventual validation of a U.S.-developed in-band solution.[75] The editors of *Radio World* observed that a worldwide DAB standard was "an agreeable thought but not necessarily an economic need." They were convinced the United States, with its affluent base of 200+ million consumers, could swing the global digital radio trajectory toward the as-yet-unproven in-band solution, and implored the broadcast industry to assume even tighter control of the technology's testing and validation process.[76] USADR and Lucent prepared more demonstrations,[77] one of which included a Lucent-sponsored "field trip" similar to the show USADR scripted in Las Vegas.[78]

The NRSC began plans in late 1995 to test all proposed digital radio systems in one U.S. market. San Francisco was chosen for its diverse terrain, in order to challenge each technology to the maximum extent regarding signal coverage and robustness.[79] There were significant difficulties securing temporary authorizations from the NTIA to establish Eureka 147 transmission facilities in the testing zone;[80] coupled with reticence from consumer electronics manufacturers that the NRSC would construe any evaluation to maximize beneficial outcomes for an in-band system, the field-testing schedule slipped into 1996.[81]

The suspicion of consumer electronics manufacturers that the NRSC's field tests would not be objective was well founded. Due to their material investment in the development of an in-band system, broadcasters had a predisposed preference, while the consumer electronics industry held out hopes that a Eureka 147-style plan for the United States could be crafted.[82] It was skeptical from the outset that a practical in-band solution was even possible, and very wary of the way USADR flew "hand-picked" groups of broadcasters to its testing sites to demonstrate its technology in a hyper-controlled environment.[83] This forced a patina of solidarity around the in-band idea. "Can a trade organization representing one faction of those who will decide the fate of DAB in the U.S. purport to perform objective tests on the one hand and support a caucus of its large members on the other hand?" asked *Radio World*'s editors. "And what happens if all systems perform well, but not exactly in a way that's comparable? Would the [consumer electronics industry] ignore broadcasters' stated position to support only [in-band DAB]?"[84]

Such criticism was not necessarily fair. The consumer electronics industry had already conducted its own independent analysis of potential in-band systems, made the process transparent to both the NAB and NRSC, and, in the end, declared them infeasible.[85] Attempts by the broadcast industry to cast doubt on the credibility of these tests were quashed in a declarative letter published in *Radio World* by those who oversaw the work and analyzed its results, including some broadcast engineers.[86] Despite the empirical evidence, in-band proponents would continue a smear campaign against the science behind the consumer electronics industry's evaluation.[87] In response to this attack, the Electronics Industry Association (EIA)[88] agreed to retest all digital radio systems, and *Radio World* observers confirmed the validity of the test protocols.[89] Described by the publication as a "dramatic hit," the EIA concluded that in-band digital broadcasting would create destructive interference to both "host" analog signals and to adjacent stations. This problem was most significant on the AM band. USADR acknowledged but downplayed these problems, and vowed that the equipment tested in the NRSC's San Francisco assessment would be dramatically improved from the units tested by the EIA.[90]

When the EIA polled its membership on raising the necessary funds to support further digital field tests, the reaction was not positive.[91] James B. Wood, president and chief engineer of Inovonics, Inc., a manufacturer of broadcast equipment in Santa Cruz, California, was blunt. "We recently received a letter in the mail . . . asking not for our ideas about digital radio, but for a contribution between $10,000 and $40,000 (whatever we might have in petty cash?) to fund the EIA and/or the NRSC evaluation of proposed systems. What does this say about digital radio in the U.S.?"[92] By the mid-1990s, a notable chasm existed between radio broadcasters and receiver manufacturers: the former was committed to an in-band digital radio solution, while the latter had not only demonstrated that such technology was undesirable, but also expressed open dismay at the idea of having to build a receiver base to support it. The lack of strategic alignment between broadcasters and receiver manufacturers over the future of U.S. digital radio would continue to resonate throughout the technology's later development and proliferation.

By the time of the annual NAB convention in April 1996, broadcasters were scratching their heads over the future of digital radio. The EIA's tests called into question the very viability of in-band systems, and the NRSC had yet to weigh in on the issue. No proponents held demonstrations or generally promoted their technology at the convention. Equipment for the NRSC's field tests had not yet even been fully built.[93] The mood was one of in-band digital radio facing "its darkest hour." Existing test results were openly termed "disappointing." Yet proponents stood up in panel discussions and gave impassioned speeches that the faults of in-band systems were simply part of the growing pains associated with developing any new technology, not rooted in a fundamental conflict with physics. The preferred system simply needed time to "mature."[94]

THE TELECOMMUNICATIONS ACT'S IMPACT ON DIGITAL BROADCASTING

1996 would prove to be a pivotal year, both for the developmental trajectory of U.S. digital radio and the broader radio industry. That year saw the passage of a new Telecommunications Act which, among other things, greatly deregulated ownership of the U.S. radio industry. Prior to 1996, the number of stations a single entity could own was tightly capped, at both the local market level and nationally. The new rules dramatically relaxed the number of broadcast outlets a single company could own in any given market, and abolished the national station ownership limit.[95] Within six years of the Act's passage, the number of commercial radio station owners would decline by 34%, while the number of actual stations rose by 5.4%.[96] As a result, much like the housing bubble of the last decade, prices for individual radio stations, especially in major markets, ballooned out of all proportion to their actual potential to generate revenue. As the radio industry consolidated, new conglomerates, such as Clear Channel, thrived in an environment that encouraged fast growth and rash cost-reduction measures that centralized production, cut jobs, and diminished program diversity. Meanwhile, older media companies, such as Westinghouse, looked to increase the synergy between their existing media holdings. Such conglomerates would end up controlling the majority of the industry's revenue stream: by 2002, one broadcast company controlled, on average, 40% of the advertising revenue in any given market; in 23% of Arbitron-ranked radio markets, the top two broadcasters controlled more than 80% of market revenue.[97]

Between 1996 and 2000, radio industry advertising revenue grew by nearly 54%.[98] Furthermore, a large infusion of investment capital into the industry took place as several radio conglomerates leveraged their operations on Wall Street, and capitalization of the radio industry rocketed from the hundreds of millions to hundreds of billions of dollars. Considering that many of those who had taken maximum advantage of consolidation were investors in the development of in-band digital radio broadcasting, it was inevitable that some of the largesse engendered by consolidation would find its way into digital development. The new Telecom Act also mandated the FCC to assume a panoply of new oversight and enforcement responsibilities that extended far beyond broadcasting, and the agency was directed to favor economic metrics over all others in the pursuit of future regulation. Burdened by an increased workload in which broadcasting assumed a lower priority, and faced with a radio industry undergoing a historically unprecedented increase in economic clout, the FCC effectively ceded any meaningful involvement in digital radio's development to the new industry behemoths.

USA Digital Radio represented the "insider" proponent of radio's digital future. As consolidation placed the fiscal fulcrum of the industry more firmly in the hands of conglomerates, and considering that USADR was the child of such firms, the industry's largest commercial broadcasters lined up to invest in the venture following the passage of the Telecom Act. During 1996, Westinghouse acquired CBS; CBS announced plans to acquire Infinity Broadcasting; and the Gannett Company sold its radio assets. All of the original remaining "Project Acorn" investors were now consolidated under one corporate roof. Westinghouse placed USADR under the direction of its Baltimore-based Wireless Solutions division and appointed Robert Struble to oversee the effort.[99] According to Struble, until this reorganization, the corporate perspective behind in-band digital radio development had been had been one of "a big science project . . . it wasn't really a whole-hog effort."[100] Within a year of the Telecom Act's passage, USADR had lined up 15 of the 20 largest radio conglomerates as investors in its venture, which collectively controlled "more than two thousand radio stations, served thousands of other affiliate stations, beamed signals to a potential audience of more than 110 million people, and took in nearly half of all radio industry revenues in the United States."[101]

USA Digital Radio engineers presented a paper at the 1996 NAB Radio Show outlining an "improved" in-band system, which would theoretically provide "virtual CD-quality stereo audio" for FM broadcasters as well as capacity for ancillary datacasting. USADR's AM system would deliver audio "quality comparable to present analog FM," along with a trickle of datacasting overhead. The company's engineers predicted they would finish their simulations of the improved system by 1997.[102] Lucent and USADR also discussed the idea of combining their development efforts toward a single in-band standard; the partnership sparked for ten months but later died, due to USADR's concerns about the influence of non-broadcast constituents on the development process.[103]

Telecom Act-related consolidation also spurred conglomerate broadcasters to upgrade the physical plants of their radio stations, so as to maximize their value in the event of future sale (a process later popularly recognized in the housing bubble as "flipping"), or as part of a "clustering" process in which the operation of several radio stations was consolidated into a single location. This led to a surge in orders for analog AM and FM transmission plants, thus inadvertently pushing the short-term implications of digital development aside and signifying that the U.S. radio industry considered any digital transition to be a long-term issue economically. Enthusiasm for any form of digital radio, especially in small and medium-sized radio markets, was nearly nonexistent in the wake of the Telecom Act.[104] Regardless, the largest players in the industry had begun to line up behind USADR; rhetoric expressed in the trade press hailed the move as an historic step forward for digital radio.[105] This did not stop some, particularly in the consumer electronics industry, from imploring that study of Eureka 147 should continue.[106]

At an NAB-sponsored engineering conference in 1996, all eyes and ears were focused on the progress of in-band system research. Attendees generally agreed that the interference issues were "probably . . . solvable." The NAB blamed pessimism about its chosen technology on "press coverage," which did not exist outside of the trades. FCC staff finally held two meetings with members of the USADR and Lucent development teams, as well as with NAB representatives. The agency favored an industry-sponsored digital solution; proponents of alternate-band digital radio technology, such as Eureka, were greeted "somewhat skeptical[ly]." Even so, the mood among those gathered at the conference was certainly not united: "To me, [in-band digital radio development involves] a lot of solutions looking for problems," said Carlos Altgelt, a supervisor in the Automotive Components Division at Ford Motor Company. Ford product design engineer Joseph Huk put the situation in stark terms: "The only [DAB system] that is going to succeed is the one that doesn't cause interference and provides a better quality service."[107]

The summer and fall of 1996 were tortuous times for the proponents of an in-band solution. As the NRSC finalized its field-test preparations, a scuffle developed between the technology's primary developers. USADR and Lucent each provided some of the equipment to be used by the NRSC, though each company claimed that the terms under which the tests would be conducted left them at a disadvantage.[108] Further complicating matters, some of the stations in San Francisco that had originally volunteered to be guinea pigs for testing had been sold following the passage of the Telecom

Act, and the new owners wanted no further part in the process.[109] In a dramatic gesture, USADR pulled its system from the NRSC's analysis. This left only the Lucent and Eureka 147 systems to be tested in San Francisco.[110] Some commentators opined that the results would "simply yield a lot of data and no consensus on a U.S. DAB standard."[111]

NRSC field evaluation of the Lucent in-band FM digital radio system commenced in July. Philip Kayne, a *Radio World* correspondent who rode along for a part of the test, was impressed by the data collection, as well as the methodology behind the process. Kayne gave the Lucent system a thumbs-up: "I firmly believe that 'digital FM' of this quality can become the broadcast standard of the future," he reported.[112] As the results of the tests were compiled, *Radio World* scored an exclusive interview with NAB President Eddie Fritts. He thought the fieldwork was useful for testing the concept of in-band digital radio, but highlighted the fact that not all proponents had been tested. For this, he squarely blamed the consumer electronics industry, which in his view was still biased toward an alternate-band solution and had, through its members on the NRSC, "controlled [the DAB exploration process] from day one." Fritts then offered the NAB "as a secretariat to facilitate testing of in-band [DAB] in a fair and impartial, underlined, capitalized, manner."[113] Given that many major NAB members were now investors in USA Digital Radio, Fritts's commentary signaled industry intent to back the company as DAB's leading proponent, despite the lack of any meaningfully independent testing.

Before the interview, the editors of *Radio World* had tentatively proposed that the U.S. radio industry "should begin to feel a little pressure to move forward on this technology," especially considering the global adoption rate of Eureka 147 at the time.[114] Following the Fritts feature, the publication clamored for the industry to "step up and take control of the development of DAB testing for the United States . . . The focus of U.S. testing should be directed at finding an in-band system that fits the bill for U.S. radio . . . Receiver manufacturers will not ignore the U.S. consumer market; it is safe to say they will cater to it."[115] If necessary, the NAB should "commandeer the . . . testing process and ensure its fairness to all proponents involved."[116] Meanwhile, field tests of the Lucent system had to be halted because the station involved in the evaluation was sold and the new owners declined to continue the experiment.[117]

Radio receiver manufacturers crowed about the collapse of the field tests. Consumer Electronics Manufacturers Association (CEMA) president Gary Shapiro pronounced it a "failure of both system performance and real interest from broadcasters" and "calls into question the future of [in-band digital radio]."[118] Further salting the wound, Shapiro wrote a scathing commentary to *Radio World*. "So, they finally admitted it!" he wrote:

> After accusing us of all sorts of sins, including bad testing and unfairness . . . By their own admission, the USADR system [has major

performance issues] . . . These are the *same* findings shown by the EIA/ NRSC lab tests . . . USADR consistently claimed its system worked fine and that our process was flawed. However, we committed to fair and impartial testing and that is what we provided. I won't hold my breath waiting for the apology. Until USADR changes its management team and [*Radio World*] adds some healthy skepticism, I will doubt any of their future claims without further unbiased, third-party testing.[119]

In an editor's note, *Radio World* stood by its coverage. Within a year, all formal cooperation between CEMA and the NAB on the issue of digital radio outside of the auspices of the NRSC would dissolve, and CEMA's co-chair on the NRSC's DAB Subcommittee was essentially reduced to a figurehead. CEMA would try one last time to influence debate in 1998 by taking the extraordinary step of formally endorsing the Eureka 147 system for use in the United States: of all the protocols examined, "only the Eureka 147 DAB system offers the audio quality and signal robustness that listeners would expect . . . in all reception environments." CEMA concluded that in-band digital radio was "not feasible at this time due to deficient performance." Critically, however, CEMA pledged to refrain from conducting official "advocacy of any system at the request of the broadcasters who said they needed more time to correct the flaws of the [in-band] system."[120]

Although regulatory intervention could have been beneficial at this juncture to moderate the inter-industry squabbles over radio's digital future, the FCC was missing in action.

SETTING THE STAGE FOR DIGITAL RADIO POLICYMAKING

Proponents of an in-band digital radio system spent 1997 attempting to make analog and digital radio signals peacefully coexist. From the perspective of public broadcasters, the "daunting technical problems and a general lack of enthusiasm for the transition" hampered these efforts. Some implored the FCC to take a more active role. According to Don Lockett, vice president of technology for National Public Radio, "Until there's a mandated timetable, as in television, I don't see that progress happening in radio." USADR told the NAB it would have a workable system ready for peer review by mid-to-late 1998. Milford Smith, chairman of the NRSC DAB Subcommittee, frankly described the situation as dire: he characterized USADR's research as "a maximum effort and perhaps the final effort at trying to make this thing work." The company, in his view, was "basically going back to the drawing board."[121] Lucent's development efforts were not hot news; USADR's continued failure to demonstrate a viable in-band digital radio system was.

Skip Pizzi, a technology manager for Microsoft with ties to public broadcasting—who would later become a member of the NRSC and regular

Radio World commentator on digital radio—viewed the work of USADR and Lucent as "kind of like a fairy tale, technologically speaking, but seductive from a business point of view." He also warned the stakes were high: "It's never a good idea to have a business concept in place before the technology can allow it," said Pizzi. "There's no Plan B in the United States."[122] In its budget request for 1997, the Corporation for Public Broadcasting asked for $22.5 million to ready its member-stations for an eventual digital radio transition. The total subsidized cost of the transition was estimated to be $50 million, and the figures were completely based on the notion of a workable in-band solution. According to NPR's Lockett, individual stations could expect to incur "upgrade" costs between $100,000 and $125,000 to broadcast a hybrid analog/digital signal—a figure five times the lowest cost estimate first projected by *Current* in 1993. As for alternate-band solutions, Lockett suggested that "radio may face a complex simulcast transition like TV—scrambling for frequencies and paying costs three times greater" than what an in-band system would offer. "Most people don't want to talk about this yet."[123]

Despite the falling-out between broadcasters and consumer electronics manufacturers, 1998 was expected to be a fruitful year in the development of digital radio. According to *Radio World*, USADR and Lucent scientists planned to "conduct over-the-air field tests of their DAB systems by autumn. They'd like to see transmitter and receiver manufacturers begin the rollout process in 1999, followed by 10 to 15 years in which stations would . . . broadcast a hybrid [in-band DAB] system, compatible with both analog and digital receivers."[124] In February, the NRSC prepared to evaluate their proposals.[125] According to David Maxson, an NRSC member and author of the definitive engineering textbook on in-band digital radio, the committee's initial inquiry was simple: "Can [in-band DAB] be better than analog? The next question, if the answer to the first were yes, is what the impact of [DAB will be] on existing analog service; can they coexist?"[126] These were the same questions raised by NRSC chairman Charles Morgan three years earlier; they would now become the primary technical metrics on which all future digital radio system evaluations would be based, and signified a decrease in expectations from the NRSC's initial criteria.

Although the NRSC declared it would only evaluate in-band systems going forward, the dearth of actual knowledge about either USADR's or Lucent's technology precluded it from developing test specifications.[127] Instead, the NRSC would "would tell [in-band] proponents for what information that NRSC was looking, and, in a general way, how it should be accomplished."[128] This gave increased latitude for proponents to massage their data toward the objective of an endorsement. With leadership of the NRSC, and especially its DAB Subcommittee, firmly in the hands of people who worked for broadcasters invested in USADR, it was clear by this point that large broadcasters alone would make the constitutive choices about radio's digital future. The FCC said nothing about this radical shift in the nature of the NRSC's testing protocol.

The rejuvenation of the NRSC's DAB Subcommittee in 1998 spurred a flurry of new activity within the development sphere. A new competitor in the race to make an in-band system work appeared on the scene. Digital Radio Express (DRE) announced it would soon propose its own "prototype" of an FM digital radio system. According to *Radio World*, DRE's cofounder, Derek Kumar, was once a subcontractor involved in USADR's preliminary research. USADR scrapped his designs after its 1996 reorganization, and Kumar continued to work on his prototype independently.[129] In initial meetings with the NRSC, DRE's technology did not impress, though committee members said they were open to any system that demonstrated a semblance of viability. In response to the newcomer, USADR and Lucent resumed joint development work.[130] At the time, the NAB's director of advanced engineering, David Layer, characterized the NRSC's goal as simply to "evaluate [in-band] technology and determine whether it's viable or not."[131] The NRSC itself hoped it could complete its in-band digital radio system evaluations by the end of the year, with "the beginning of industry rollout by summer of 1999."[132] FCC approval of any system, it would seem, was already a foregone conclusion, despite the technical challenges that lay ahead.

As further details about the functionality of in-band DAB emerged, rank-and-file broadcast engineers began to openly question its viability. Mike Worrall, the assistant chief engineer for a cluster of stations in Los Angeles, was not impressed with the presumed audio quality of in-band digital broadcasts, and predicted that digital radio would sound worse than analog. "Why is [DAB] seen as such a necessary development for terrestrial broadcasting . . . if the audio quality *by definition* will be compromised?," he asked in *Radio World*. "I'm beginning to sense that the emperor has no clothes."[133] E. Glynn Walden, the director of engineering for CBS Radio and a man whose involvement with in-band digital radio dated back to Project Acorn, responded directly. According to Walden, USADR and Lucent's digital audio encoding algorithms were more sophisticated than garden-variety techniques such as MP3 compression; they used "perceptual audio coding," which could provide "digital CD quality within the broadcast channel."[134]

In March 1998, *Radio World* sat down with Walden for an interview about USADR's developmental work. He began by advancing the dominant argument for radio's digitalization: "The world is going digital. We are going to make [DAB] optimized for broadcasting and for our listeners. If we don't do that, we have nothing . . . We have ten years of implementation issues." In the next breath, Walden predicted digital radio stations would be on the air within a year, with receivers available to the public by 2000, and that there would be "nothing analog" left on the air shortly afterward.[135] As a kicker, Walden also revealed more details about collaboration between USADR and Lucent that suggested the two companies were discussing a merger.[136] At the NAB's annual convention in Las Vegas, USADR and DRE debated the merits of their digital radio systems in front of interested broadcasters. USADR positioned itself as the front-runner in the race to develop a workable system, and announced that field-testing on the AM and FM bands would commence on 16 stations "during the first quarter of 1999."[137]

Shortly after the convention, Lucent Technologies created a subsidiary, Lucent Digital Radio (LDR), to move its development process forward. LDR's president, Suren Pai, said the company was seeded with a "substantial" amount of capital from AT&T. Lucent had developed a significant amount of intellectual property that was critical to the basic functionality of in-band DAB; consolidating it under a single corporate structure was also expected to increase the net value of this research. Pai also announced that Lucent was "walking away from" collaboration with USADR, to develop a system that does "not use technology that it [LDR] does not currently own."[138] This included the audio encoding algorithm that Lucent had developed, which USADR had already announced it would use in its own system. Lucent would now keep the codec for itself, forcing USADR to develop its own—unless it desired to license Lucent's.[139]

The creation of Lucent Digital Radio had multiple objectives. Lucent attempted to corner the market on important components of in-band DAB technology; if LDR could not win the overall development race, at least it could force USADR to share the wealth by controlling intellectual property necessary for the system's functionality. By taking its expertise in digital audio compression methods away from USADR, LDR forced its competitor to reallocate developmental resources, presumably to slow down USADR's research and give LDR a chance to design its own fully integrated AM and FM in-band system. Pai predicted that his technology would be on the air by the turn of the century.[140]

The disentanglement of USADR and LDR's joint research introduced further complications into the NRSC's proposed testing and verification of in-band systems. Most importantly, opined *Radio World*, nobody had yet addressed a fundamental question: can an in-band digital radio signal provide better quality than legacy analog service without harming it?[141]

During the summer of 1998, each proprietor jockeyed within the trade press for the position of perceived frontrunner. USA Digital Radio unveiled expanded research and development facilities in Columbia, Maryland, and announced it had commenced limited field tests of its FM system. It expected its technology would hit the market by 2000 and the digital radio transition would be completed within ten years.[142] Digital Radio Express reported that it had a volunteer station on which to field-test its system, and an experimental license from the FCC to conduct the work.[143] By September, USADR and DRE had completed limited field tests of their systems, while LDR claimed to be not far behind. Meanwhile, the NRSC announced it hoped to have a package of formal testing guidelines drafted and ready for industry comment by the end of the year.

For their part, FCC staff said they would expedite the processing of applications for experimental authorization to conduct hybrid analog/

digital broadcast tests, but otherwise would passively observe the NRSC evaluation.[144] *Radio World* reported that broadcasters would have welcomed proactive FCC involvement: according to members of the NRSC's DAB Subcommittee, there was difficulty in settling on test criteria, as all three proponents kept significant technical details of their systems "close-to-the-vest." According to the NAB's David Layer, "As an engineer, I find it hard to get satisfaction from the proponents in terms of what they're doing. But I'm just going to have to deal with that, because from a business standpoint, they're doing the right thing." Due to this complication, the NRSC had already agreed to farm out its actual testing to third-party laboratories; Layer said this move was inevitable because, "Frankly, if the committee's work is going to be accepted, the committee needs to be able to say more than, 'The proponents handed us this data and we looked it over.' "[145] *Radio World* later reported that because the development of NRSC consensus on testing protocols remained "painstakingly slow," proponents were preparing to submit their designs directly to the FCC for review, if necessary.[146] The FCC made no comment on these overtures and expressed no interest in deeper engagement with the NRSC process.

By the fall of 1998, the business model of in-band digital radio proponents was clear. Lucent Digital Radio had already hinted that its corporate mission was to fast-track proprietary applications to market, but not necessarily manufacture them. Similarly, USA Digital Radio and Digital Radio Express announced they would license their technology for manufacture. USADR's Robert Struble was the most explicit about the digital radio development race being one not of technological progress, but of intellectual property: "[The] business model is actually real attractive. We don't have to build the plants, we don't have to hire millions of people to go do this stuff. We just need to make sure the technology works . . . and then go out and strike some deals with these folks."[147]

USA Digital Radio would force the development issue on October 7, 1998, when it filed a Petition for Rulemaking with the FCC, asking the agency to begin the certification process for its technology. Simultaneously, USADR tendered several filings with the U.S. Patent and Trademark Office covering terminologies to describe in-band digital radio that it claimed to have invented. USADR predicted the FCC would take 18 to 24 months to act on its petition. Meanwhile, the NRSC had not yet formalized the terms under which it would evaluate the three in-band contenders.[148] Unsurprisingly, USADR's competitors were critical of its FCC petition. Lucent Digital Radio was dismissive, while Digital Radio Express called the filing "premature." DRE founder Derek Kumar said USADR "in effect, circumvented input from representatives of interested parties taking part [in the NRSC process]," and he threatened lawsuits if USADR received the patents and trademarks it desired.[149]

Radio World deemed the petition a mixed blessing. It hinted at a bumpy road ahead for the creation of a single, nationwide digital radio standard.

The publication asked all three proprietors about the possibility of future collaboration, and their responses were characterized as "lukewarm."[150] This prompted *Radio World*'s editor Paul McLane to pen a provocative plea for the FCC to take a greater role in the development process:

> We have talked seriously about DAB for eight years or so, and the process has seen numerous setbacks and detours . . . USADR clearly wants to assert "ownership" of the [in-band] DAB issue . . . The challenges for USADR now are to justify the content of its filing, to show that it has not circumvented the NRSC process, to show fellow broadcasters that the plans of this CBS-backed enterprise are consistent with the goals of all radio owners. And, of course, to show that the USADR system actually works . . . The people who run Lucent know that they could be seen as outsiders, an extension of the phone company . . . Among their challenges now is to demonstrate that their commitment is real . . . that they understand the financial and technical needs of broadcasters. And, of course, that the Lucent system actually works . . . The challenges for DRE are to communicate its goals to the industry, to demonstrate that it has the financial resources to see this project through over the next several years, to dispel its image of a minor player. And, of course, to show that the DRE system really works.[151]

The implicit suggestion made by McLane was that the FCC should work to foster collaboration between DAB developers, instead of letting the marketplace create separate standards that might be implemented at the discretion of any given station. Such an outcome could balkanize the adoption of digital radio technology, dooming it in a manner similar to the adoption of multiple analog AM stereo standards in the 1980s. He noted that FCC intervention was required to force a "grand alliance" between digital television technology developers, which ultimately led to industry consensus behind a single standard.

The FCC remained silent on the salient points of McLane's commentary. It ignored or declined several opportunities to involve itself in digital radio's development, and by accepting USADR's Petition for Rulemaking in the absence of strong technical rationales, the agency effectively signaled that it would go along with whatever digital broadcast solution the radio industry could come up with. At the very least, the FCC's acquiescence to market actors proceeding with the promulgation of a technology they did not fully understand represents a significant abdication of regulatory responsibility with regard to the agency's trustee-function involving the integrity of the public airwaves.

With the FCC on the sidelines, the NRSC's DAB Subcommittee became the forum-of-choice for the technical vetting of any system, and, over time, broadcasters with direct ties to USADR assumed key positions within the NRSC. In retrospect, USADR's petition can be seen as insurance that formalized consideration of digital radio would continue even if the NRSC's own evaluative process broke down, as it did in 1996. While consumer electronics manufacturers would continue to take part in NRSC proceedings, broadcasters effectively controlled the direction of the DAB Subcommittee's work. The effects of this rift would take years to manifest itself, to the benefit of no one.

The U.S. public radio system also played an important role in facilitating USADR's quest for primacy. National Public Radio provided all of the station-platforms on which the technology was initially demonstrated, a commitment to its earlier declaration that it would not be shunted aside in the dialogue over digital radio's constitutive choices. Although NPR and the Corporation for Public Broadcasting also did not fully understand the technical implications of an in-band system, leadership at these organizations made the strategic decision to align themselves with the will of the commercial radio industry, and ultimately with USADR.

The move by USADR to jump-start a policy discussion before any technology had been proven to even be marginally functional raised alarm in some quarters. Independent broadcasters openly questioned the viability of any in-band system and were unimpressed by the stated goal of adopting something that only provided a level of service "better than analog." Unfortunately, these concerns were swept aside by in-band proponents, aided by a trade press that reinforced notions that digital radio was inevitable and cast early critics of the technology as uninformed or seeking to interfere with "progress."

Consolidation within the radio industry, fomented by the 1996 Telecom Act, shifted a large segment of its economic power behind a single in-band proponent; the combined might of USA Digital Radio's supporters would ultimately provide enough justification to effectively preclude all but its protocol as the digital future of U.S. radio broadcasting. Importantly, this occurred before the FCC even became actively involved in the issue. Consensus among those anointed as "players" in the digital radio debate revolved around an inarticulate sense of urgency and in the absence of a clear end goal. As Chapter 3 will illustrate, placing faith in what was still essentially a concept technology would lead to potentially dangerous consequences for the future of both analog and digital broadcasting.

NOTES

1. Ken G. Pohlmann, *Principles of Digital Audio, 5th Edition* (New York: McGraw-Hill, 2005), 636.
2. See Federal Communications Commission, Docket No. 90–357, 12 FCC Rcd at 5756 (1990); and Ernest A. Hakanen, "On Autopilot Inside the Beltway: Organizational Failure, the Doctrine of Localism, and the Case of Digital Audio Broadcasting," *Telematics and Informatics* 12, no. 1 (1995), 16–17.

3. Richard Barbieri, "New Radio Sound Under Study," *Current* IX, no. 17 (September 24, 1990), 4.
4. Michael Starling, "Public Radio's Eye on the Future," *Current* IX, no. 23 (December 17, 1990), 20.
5. John Wilner, "A Little DAB Will Do Ya," *Current* XI, no. 2 (February 3, 1992), 3.
6. Pohlmann, 647.
7. See Michael P. McCauley, "Radio's Digital Future: Preserving the Public Interest in the Age of New Media," in Michele Hilmes and Jason Loviglio, eds, *Radio Reader: Essays in the Cultural History of Radio* (New York: Routledge, 2002), 508; and Richard Rudin, "The Development of DAB Digital Radio in the UK: The Battle for Control of a New Technology in an Old Medium," *Convergence: The International Journal of Research into New Media Technologies* 12, no. 2 (2006), 165.
8. Pohlmann, 643, 647.
9. David P. Maxson, *The IBOC Handbook: Understanding HD Radio Technology* (Burlington, MA: Focal Press, 2007), 16.
10. Alan Carter, "Eureka 147 Wins Nod; In-Band Not Out Yet," *Radio World*, December 28, 1994, 1.
11. See Pohlmann, 643; Alan Carter, "Europeans Move Ahead with Eureka," *Radio World*, October 4, 1995, 21; and Ted Tait, "BBC Switches on National DAB Service," *Radio World*, November 1, 1995, 7.
12. See Mary Ann Seidler, "DAB Transmitters Get a Closer Look," *Radio World*, February 8, 1995, 8; Carter, "Europeans Move Ahead with Eureka," 21; T. Carter Ross, "Eureka Plows Forward in Europe, World," *Radio World*, December 27, 1995, 11; Ronald K. Jurgen, "Broadcasting With Digital Audio," *IEEE Spectrum*, March 1996, 56; and Bev Marks, "England Begins DAB Multiplex Experiment," *Radio World*, April 3, 1996, 14.
13. See James Careless, "Pioneer Commits to DAB," *Radio World*, January 12, 1994, 3, 9; and James Careless, "Pushing DAB Envelope in Canada," *Radio World*, February 8, 1995, 6, 11.
14. See Brian O'Neill, "Digital Audio Broadcasting in Canada: Technology and Policy in the Transition to Digital Radio," *Canadian Journal of Communication* 32, no. 1 (2007), 74–75; and Jurgen, 58.
15. See O'Neill, 75–77; and James Careless, "L-Band DAB Ancillary Service Introduced," *Radio World*, April 5, 1995, 6.
16. See Petition for Rulemaking filed by USA Digital Radio Partners, L.P., RM-9395, October 7, 1998, 16-17; and David Sedman, "Radio Transmission," in Peter B. Seel and August E. Grant, eds, *Broadcast Technology Update: Production and Transmission* (Boston: Focal Press, 1997), 163.
17. Marguerite Clark, "Eureka 147 Continues To Spread," *Radio World*, July 10, 1996, 1.
18. Pohlmann, 646.
19. See Jack Robertiello, "The Competition," *Current* X, no. 1 (January 21, 1991), 8; and Jack Robertiello, "The Competition," *Current* X, no. 8 (April 29, 1991). 16.
20. Marko Ala-Fossi and Alan G. Stavitsky, "Understanding IBOC: Digital Technology of Analog Economics," *Journal of Radio Studies* 10, no. 1 (2003), 69.
21. Richard Barbieri, "Bill Would Free Spectrum," *Current* IX, no. 15 (August 20, 1990), 1, 17–18. The bill never made it out of committee.
22. Pohlmann, 647.
23. Federal Communications Commission, *Notice of Proposed Rulemaking and Further Notice of Inquiry*, 7 FCC Rcd 7776, 7778 (1992).

24. Alan Haber, "FCC Lays Groundwork for Satellite Radio," *Radio World*, February 8,1995, 1, 12–13.
25. Alan Haber, "NAB Nixes SAT DARS," *Radio World*, February 8, 1995, 12.
26. Editorial, "Satellite DAR: A Call to Action," *Radio World*, February 8, 1995, 5.
27. See McCauley, 509; and Ala-Fossi and Stavitsky, 66–67.
28. See Ala-Fossi and Stavitsky, 64, 68; and Maxson, 31.
29. Ala-Fossi and Stavitsky, 69.
30. McCauley, 509.
31. See ibid.; Skip Pizzi, "Digital Conversion Costs Becoming Clearer in Radio," *Current* XII, no. 3 (July 12, 1993), 15; Lynn Meadows, "Eureka Fights Odds for Inroad to U.S. Market," *Radio World*, December 13, 1995, 6; and Steve Behrens, "Race for Digital Radio is Uphill From Here," *Current* XVI, no. 5 (March 31, 1997), 14.
32. Steve Behrens, "Sky Not Expected to Fall When Radio Goes Digital," *Current* XII, no. 18 (October 4, 1993), 1.
33. Richard Barbieri, "Digital to Bring New World to TV, Radio," *Current* IX, no. 18 (October 8, 1990), 12.
34. Jurgen, 52.
35. See Jack Robertiello, "The Competition," *Current* X. no. 8 (April 29, 1991), 16; and Pohlmann, 660.
36. Jurgen, 57.
37. Reply Comments of iBiquity Digital Corporation, 99–325, January 24, 2012, 6.
38. John Anderson, "Digital Radio in the United States: Privatization of the Public Airwaves?" *Southern Review: Communication, Politics, and Culture* 39, no. 2 (2006), 7.
39. Brenda J. Cronin, "CPB to Study Digital Radio," *Current* X, no. 6 (April 1, 1991), 3.
40. Wilner, "A Little DAB Will Do Ya," 3.
41. Michael Starling, "What Digital Transmission Could Mean for Public Radio," *Current* X, no. 6 (April 1, 1991), 10.
42. Behrens, "Sky Not Expected to Fall When Radio Goes Digital," 1, 21.
43. Mark Starowicz, "Will Radio Survive the Digital Revolution? (By the Way, What Is Radio Anymore?)," *Current* XVII, no. 13 (July 27, 1998), 15–17.
44. Maxson, 13.
45. According to FCC records, WILL-FM's effective radiated power is 105,000 watts, which would make the transmission power of its first experimental digital broadcast at approximately 105 watts.
46. Steve Behrens, "That Caused Everyone to Say, 'This is Pretty Slick!,'" *Current* XI, no. 23 (December 14, 1992), 10.
47. Anderson, 7.
48. Leslie Stimson, "Suren Pai: Lucent Is For Real," *Radio World*, June 24, 1998, 1, 8.
49. Jurgen, 57.
50. Maxson, 15.
51. Pohlmann, 650.
52. Jurgen, 59.
53. See ibid., 56–57; Pohlmann, 651; and Behrens, "Race for Digital Radio Is Uphill From Here," 14.
54. John Gatski, "End of 1995 Likely Timetable For Completion of DAR Tests," *Radio World*, March 8, 1995, 3.
55. Nancy Reist, "A Glance at the Future of News Broadcasts," *Radio World*, November 16, 1994, 23, 27.
56. "DAB Testing Continues in U.S.—World Moves Closer to a Standard," *Radio World*, December 28, 1994, 7.

57. Judith Gross, "In-Band DAB Keeps Moving Forward," *Radio World*, November 16, 1994, 13.
58. John Abel, "Understanding the Many Possibilities of Digital," *Radio World*, February 8, 1995, 5.
59. Ibid.
60. Thomas R. McGinley, "USA Digital Radio Takes Vegas by Storm," *Radio World*, May 17, 1995, 1. By this point, KUNV was broadcasting a digital signal at less than 5% of its effective radiated analog power (700W digital compared to 15 kW analog ERP).
61. Alan Carter, "USA Digital to Headline in 'Vegas," *Radio World*, March 8, 1995, 6.
62. McGinley, 14.
63. Alan Carter and Charles Taylor, "USA Digital Radio Proves Itself at NAB '95," *Radio World*, May 3, 1995, 6.
64. Thomas Pear, "All-Digital Future within Radio's Reach," *Radio World*, May 17, 1995, 13.
65. John Marino, "NAB Boasts Varied Technical Agenda," *Radio World*, March 22, 1995, 61.
66. Pear, 13–14.
67. Ibid., p. 14.
68. Ibid.
69. See "CCA to Go Digital," *Radio World*, April 5, 1995, 48; and Carter and Taylor, 6.
70. Carter and Taylor, 1.
71. McGinley, 8.
72. Carter and Taylor, 6.
73. Mark F. McNeil, "IBOC Listening," *Radio World*, June 14, 1995, 5.
74. Robert C. Tariso, "IBOC Questions," *Radio World*, August 23, 1995, 5.
75. See Lucia Cobo, "L-Band in Doubt for Eureka Field Tests," *Radio World*, June 14, 1995, 1, 6; and Lucia Cobo, "Politics Mires Digital Audio Radio Testing," *Radio World*, June 14, 1995, 4.
76. See Editorial, "On From The Tests," *Radio World*, June 28, 1995, 5; Editorial, "In-band: The Only Choice," *Radio World*, September 20, 1995, 5; and Editorial, "Too Many Vested Interests," *Radio World*, November 29, 1995, 5.
77. Lucia Cobo, "Proponents Plan for Fall Schedule," *Radio World*, June 14, 1995, 6.
78. Lucia Cobo, "AT&T Further Proves In-Band Can Be Done," *Radio World*, October 4, 1995, 4.
79. See Lynn Meadows, "DAR Field Tests to Begin in San Francisco," *Radio World*, July 12, 1995, 6; and Lynn Meadows, "DAB Field Tests Good to Go in San Francisco," *Radio World*, January 24, 1996, 1, 12.
80. See "Eureka 147 to Test in San Francisco," *Radio World*, August 23, 1995, 1, 2; Lynn Meadows, "Proponents Create DAB Delays for U.S.," *Radio World*, November 29, 1995, 1, 8; and "DAB Tests Get L-Band," *Radio World*, January 10, 1996, 1, 6.
81. See Lynn Meadows, "EIA Responds to DAB Proponents' Concerns," *Radio World*, October 18, 1995, 1, 3; "U.S. DAB on the Slow Track," *Radio World*, December 27, 1995, 3; and "Answers Wanted On DAB," *Radio World,* February 21, 1996, 1, 12.
82. Judith Gross, "Group Heads Impressed by DAB Demo," *Radio World*, October 19, 1994, 10–11.
83. Ibid.
84. Ibid., 11.

85. See Judith Gross, "EIA DAB Tests Viewed as 'Unfair,' " " *Radio World*, December 14, 1994, 6, and Judith Gross, "RF Mask Not So Strict," *Radio World*, December 14, 1994, 6.

86. Randall T. Brunts and Alfred Resnick, "Digital Radio Testing Fair," *Radio World*, January 11, 1995, 5.

87. See Lucia Cobo, "In-band Slammed by EIA Testing Process," *Radio World*, September 20, 1995, 1, 3; Editorial, "In-band: The Only Choice," *Radio World*, September 20, 1995, 5; Lucia Cobo, "EIA Testing Process Needs Some Work," *Radio World*, September 20, 1995, 4; and "DAB Lab Tests," *Radio World*, October 4, 1995, 2.

88. Later to become the Consumer Electronics Manufacturers Association (CEMA) in 1996–97, and the Consumer Electronics Association (CEA) in 1999.

89. "IBOCs to Be Retested," *Radio World*, April 19, 1995, 3; and Lucia Cobo, "Digital Radio: Just Rocket Science?" *Radio World*, May 31, 1995, 4.

90. Lynn Meadows, "Tests Show AM Digital Problems," *Radio World*, March 20, 1996, 1, 6.

91. "Money Sought for DAB Testing," *Radio World*, January 12, 1994, 2, 9.

92. James B Wood, "Mailbag Musings," *Radio World*, January 12, 1994, 5.

93. Lynn Meadows, "NAB '96: Waiting for U.S. DAB Standard," *Radio World*, March 20, 1996, 9–10.

94. Lucia Cobo, "Thinking About DAB and the Road Left to Travel," *Radio World*, May 15, 1996, 4.

95. See Eric Klinenberg, *Fighting for Air: The Battle to Control America's Media* (New York: Metropolitan Books, 2007), 26–27; and Robert W. McChesney, *The Political Economy of Media: Enduring Issues, Emerging Dilemmas* (New York: Monthly Review Press, 2008), 419.

96. George Williams and Scott Roberts, *Radio Industry Review 2002: Trends in Ownership, Format, and Finance* (Washington, D.C.: Federal Communications Commission Media Ownership Working Group, 2002), http://www.fcc.gov/ownership/materials/already-released/radioreview090002.pdf, Executive Summary, 3.

97. Federal Communications Commission Media Ownership Working Group, Main Report, 6.

98. Mark Fratrick, "Where Is the Radio Industry Going," BIA Financial Network, 2006, http://www.bridgeratings.com/bia_radioanalysis_2006.pdf, 2.

99. "USADR Proceeds With DAB Plans," *Radio World*, November 13, 1996, 12.

100. Quoted in Eric Rhoads, "Robert Struble: At Long Last . . . Digital," Radio Ink, September 23, 2002, http://www.radioink.com/listingsEntry.asp? ID=90913 &PT=industryqa.

101. McCauley, 510.

102. Lucia Cobo, "Good News for IBOC AM and FM," *Radio World*, October 30, 1996, 4.

103. See Lynn Meadows, "AT&T Pulls IBOC from DAB Tests," *Radio World*, October 16, 1996, 11; James Careless, "Canada Poises for DAB; U.S. Thinks About It," *Radio World*, December 25, 1996, 1, 11; and Anderson, 7.

104. See Alan Haber, "Transmitter Manufacturers Eye DAB Progress," *Radio World*, September 4, 1996, 19; and Sedman, 164.

105. See Editorial, "The Time Is Now," *Radio World*, March 6, 1996, 5; and Editorial, "DAB: Stay the Course," *Radio World*, May 15, 1996, 5.

106. Scott Wright, "A Eureka 147 Solution For the United States," *Radio World*, May 15, 1996, 5, 14.

107. Lynn Meadows, "On the Slow Track with U.S. DAB," *Radio World*, May 29, 1996, 17.

108. See Lynn Meadows, "DAB Field Tests Inch Closer to Start Date," *Radio World*, May 15, 1996, 1, 10; and Lynn Meadows, "Tests to Start Regardless," *Radio World*, June 12, 1996, 1, 11.
109. See Meadows, "Tests to Start Regardless," 11; Sedman, 162; and Anderson, 8.
110. Meadows, "Tests to Start Regardless," 11.
111. Lynn Meadows, "Tests Begin in San Francisco," *Radio World*, August 7, 1996, 1, 13.
112. Philip M. Kane, "Day in the Life of DAB Field Tests," *Radio World*, October 2, 1996, 1, 3.
113. Lucia Cobo and Alan Haber, "NAB Poised to Push IBOC," *Radio World*, October 2, 1996, 55.
114. Lucia Cobo, "Still Looking for DAB Answers," *Radio World*, September 18, 1996, 4.
115. Editorial, "Stop Wasting Time," *Radio World*, October 2, 1996, 5.
116. Lucia Cobo, "NAB Readies Itself for IBOC Action," *Radio World*, October 2, 1996, 4.
117. Lynn Meadows, "AT&T Pulls IBOC from DAB Tests," *Radio World*, October 16, 1996, 1, 11.
118. Ibid., 11.
119. Gary Shapiro, "Not Holding His Breath," *Radio World*, December 11, 1996, 5.
120. See Jurgen, 53; and Anderson, 9.
121. Behrens, "Race for Digital Radio Is Uphill From Here," 1, 14.
122. Ibid.
123. Steve Behrens, "Sought: 45% of DTV Cost," *Current*, October 6, 1997, http://www.current.org/wp-content/themes/current/archive-site/tech/tech718d.html.
124. Editorial, "DAB Heats Up," *Radio World*, February 18, 1998, 5.
125. Bob Rusk, "NRSC Reactivates DAB Committee," *Radio World*, February 18, 1998, 10.
126. Maxson, 36.
127. See Maxson, 38; and Rusk, "NRSC Reactivates DAB Committee," 10.
128. Maxson, 38.
129. See Bob Rusk, "A New Player Enters IBOC Effort," *Radio World*, February 18, 1998, 1, 21; and Thomas R. McGinley, "AAC Finds New Fans at DRE," *Radio World*, March 18, 1998, 21–22.
130. Leslie Stimson, "NRSC Group Meets New DAB Player," *Radio World*, March 6, 1998, 1, 8.
131. Ibid., 8.
132. Ibid.
133. Mike Worrall, "Emperor's Clothes," *Radio World*, February 4, 1998, 5.
134. Glynn Walden, "IBOC Delivers," *Radio World*, February 18, 1998, 5.
135. Paul J. McLane and Leslie Stimson, "Walden: 'This Is the Future of Radio,'" *Radio World*, March 18, 1998, 17.
136. Ibid., 19.
137. Leslie Stimson, "IBOC DAB Proponents Square Off," *Radio World*, April 29, 1998, 1, 6.
138. Leslie Stimson, "Lucent Enters IBOC Fray Alone," *Radio World*, May 13, 1998, 1.
139. Stimson, "Suren Pai: Lucent Is For Real," 8.
140. Ibid., 10.
141. Leslie Stimson, "In-Band, On-Channel: Now What?" *Radio World*, June 10, 1998, 19.
142. See Leslie Stimson and Rick Barnes, "USADR Facility Is Unveiled," *Radio World*, July 8, 1998, 1, 19; and Leslie Stimson, "USADR Conducts Multipath Tests," *Radio World*, August 5, 1998, 8.

143. Leslie Stimson, "DRE Ready for an Early Test," *Radio World*, July 22, 1998, 3.
144. Leslie Stimson, "IBOC DAB Process Moves Ahead," *Radio World*, September 16, 1998, 1, 16.
145. Leslie Stimson, "Where Is IBOC DAB Going?," *Radio World*, September 30, 1998, 14.
146. Ibid., 1, 14, 17.
147. See Leslie Stimson, "Struble Pilots USADR IBOC Ship," *Radio World*, August 5, 1998, 8, 12; and Stimson, "DRE Ready for an Early Test," 3.
148. Leslie Stimson, "USADR to File 'Historic' Petition," *Radio World*, October 14, 1998, 1, 12.
149. See Leslie Stimson, "USADR Filing Ignites Debate," *Radio World*, November 11, 1998, 1, 8; and Leslie Stimson, "IBOC Technical Details Emerge," *Radio World*, November 11, 1998, 3, 7.
150. Stimson, "IBOC Technical Details Emerge," 7.
151. Paul J. McLane, "From the Editor: Seattle and the 'DAB' Radio Show," *Radio World*, November 11, 1998, 4.

3 The Fundamental Deficiencies of IBOC DAB

According to the National Radio Systems Committee, a successful digital radio technology is one that provides "a significant improvement over the analog systems currently in use."[1] To understand the chosen U.S. digital radio system's troubles with regard to achieving this goal requires an overview of its fundamental deficiencies. The system's invasive spectral occupancy, lack of meaningful bandwidth capacity, and wholly proprietary nature conspire to keep it from reaching a threshold of transformative potential for radio broadcasting.

SPECTRAL OCCUPANCY

In Chapter 2, two general categories of digital radio broadcast technologies were referred to, for the sake of simplicity, as "in-band" and "alternate-band" digital audio broadcasting. The United States' digital radio protocol is called "In-Band-On-Channel," or IBOC, DAB. The name implies that analog and digital audio signals peacefully reside on the same swath of spectrum. This is untrue.

Traditionally, the FCC has allowed analog FM stations to occupy approximately 200 kilohertz (kHz) of spectrum; this is why the FM dial is segmented into 200 kHz channels. Figure 3.1 clearly shows that a hybrid analog/digital FM-IBOC signal actually occupies nearly 400 kHz of spectrum, effectively appropriating new spectrum to broadcast the digital portion of the transmission. In the Figure 3.1 schematic, the triangle represents an FM station's analog signal; the squares on either side represent the digital "sidebands." An "extended hybrid" mode allows stations to widen their digital sidebands toward the center of their analog signal, thus sacrificing some analog signal quality for more digital capacity, if desired.[2] These digital sidebands carry duplicate information in case one of them is interfered with by an adjacent channel signal.[3]

If the FCC's FM allocation rules require a spacing between stations of at least 200 kHz, and an FM-IBOC hybrid analog/digital signal occupies

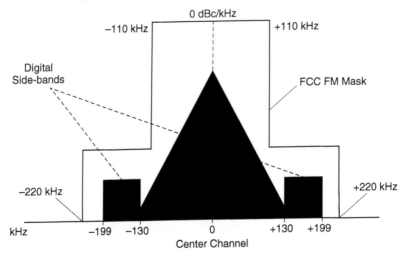

Figure 3.1   Waveform schematic of a hybrid analog-digital FM-IBOC signal.[4]

twice that space, how is IBOC allocation permissible under current broadcast rules? Looking again at Figure 3.1, the entire FM-IBOC waveform is outlined by what is described as the "FCC FM Mask." Historically, the FCC has required emission masks for analog transmissions;[5] these were initially intended to protect broadcast stations from potential interference generated by their neighbors on the dial. For example, sometimes a transmitter malfunctions, and this can cause off-frequency emissions to occur. Or a station may overmodulate its signal, thereby "splattering" its transmission onto adjacent frequencies.

Proponents of IBOC have cleverly appropriated the spectrum covered by the FCC's emissions mask around each station as the station's own.[6] This is a fundamental reinterpretation of the mask's regulatory intent. In simple terms, the FCC rule was designed to provide a "guard band" of fallow spectrum between stations to protect against *spurious and transient* emission—*not* to be utilized for the continuous transmission of energy. This historical fact seems to be lost on the agency itself.[7] In fact, the NRSC convinced the FCC to expand the amount of allowable energy that may reside within the emissions mask in order to accommodate IBOC signals—otherwise some of the "spurious noise" generated by the system would have been legally impermissible.[8]

IBOC proponents claim that such a fattened spectral footprint is necessary to accommodate both analog and digital radio signals. However, the all-digital FM broadcast mode, illustrated in Figure 3.2, shows the aggregate signal footprint remains unchanged—still double that of an analog FM signal.[9] Space previously occupied by the analog signal is simply repurposed for additional digital capacity.

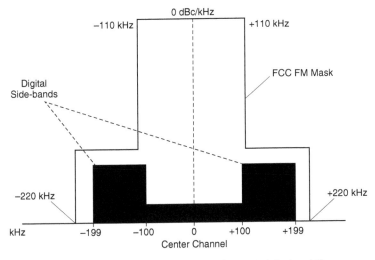

Figure 3.2   Waveform schematic of an all-digital FM-IBOC signal.[10]

Therefore, stations that transition to digital broadcasting will use an increased amount of spectrum *in perpetuity*. This has caused many engineers, including NPR's Mike Starling, to urge the technology's redefinition as "in-band-adjacent-channel," or IBAC, since that is what it actually represents.[11]

The fattening of FM broadcast signals creates the real potential for analog-to-digital interference, which IBOC's proprietors acknowledge is an ongoing concern.[12]

Figure 3.3 represents a scenario in which two FM stations on adjacent channels, both operating with IBOC sidebands, might cause destructive interference to each other.

Not only do portions of their analog signals overlap, but the digital sidebands of both stations also directly impinge upon each other's analog transmissions, which in the real world sounds like a buzzing noise on analog radio receivers. Listeners do not understand the nature of the interference caused by digital sidebands and, therefore, may believe there is something wrong with their receivers, when the problem is actually due to the imposition of new radio frequency (RF) energy on the radio dial that the receivers were never designed to accommodate.[13]

Furthermore, the danger of self-interference exists between the analog and digital portions of a single FM-IBOC signal. Considering that the digital sidebands overlap the edges of a station's analog "host" transmission, the potential exists for digital data to leak through the filters of analog radio receivers and cause interference that sounds like white noise. There is no concrete way to prevent this potential self-interference, whose likelihood increases as stations adopt the extended hybrid broadcast mode—a mode never thoroughly evaluated by the NRSC.[14]

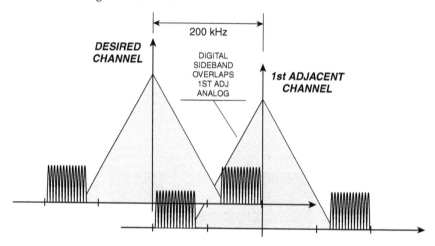

Figure 3.3   Two FM-IBOC stations on adjacent channels with overlapping analog and digital signals.[15]

Due to the limitations of broadcasting within the emissions mask, the digital sidebands of an FM-IBOC signal must be broadcast at a power level equivalent to between 1% and 10% of a station's analog transmission. Thus, digital radio signals cannot replicate the full analog coverage area of any given station.[16] Penetration of digital signals in buildings is extremely poor and, in some cases, described as "impossible."[17] According to National Public Radio, "Resurgence in the use of outdoor antennas could be promoted" to address this problem, but the majority of radio listeners are not expected to spend the time and effort to install them.[18] This especially worries public broadcasters, as half of an average public radio station's revenue is based on listener contributions, and any broadcast technology that degrades a station's coverage area represents a threat to its fiscal health.[19] Despite the weak nature of digital sidebands, transmitter manufacturers caution adopting stations to budget for increased electricity and cooling costs at their transmitter plants when they broadcast in digital—perhaps by as much as 100%.[20]

IBOC's proprietors have developed a workaround to the problem of diminished digital coverage. When a digital signal begins to degrade, the radio receiver "gracefully" blends to the analog signal of the same station, thereby avoiding the "cliff effect" common to other digital audio platforms.[21] Proponents of the technology qualify the blend-to-analog function as a mechanism to guarantee that an "IBOC signal can never be worse than, and is usually much better than, the performance afforded by existing analog service."[22] This is a questionable assertion, given that digital sidebands can also cause harm to the quality reception of analog signals.

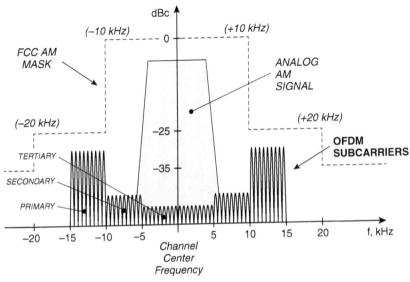

Figure 3.4   Waveform schematic of a hybrid analog/digital AM-IBOC radio signal.[23]

On the AM side of the technology, the potential for interference and diminished coverage is much worse. According to David Maxson, the basic nature of the FCC's analog AM station allocation rules exacerbates this situation.[24] While an FM station is allocated 200 kHz per channel, an AM station must get by with just 5% of that spectral capacity, or 10 kHz per channel. AM-IBOC also appropriates the FCC's emissions mask under which digital information is placed. In fact, one of the technology's developers has described this practice as "an expansion of the use of the AM band."[25]

As shown in Figure 3.4, an AM-IBOC hybrid radio signal actually occupies 30 kHz—triple the analog footprint.

Unlike hybrid FM signals, AM-IBOC transmissions place digital data directly underneath the analog signal, though since its introduction new broadcast modes have been developed that allow stations to avoid this mixing of analog and digital signals.[26] If an AM station does opt to include underlying digital data, the analog signal must be transmitted monophonically.[27] Listeners may notice this part of the digital transmission as a hiss on their receivers. If and when AM stations go all-digital, as shown in Figure 3.5, their spectral footprints will decrease, but still remain double that of the analog system.[28]

AM-IBOC carries with it a plethora of interference concerns. AM signals travel in two ways: through the ground (these are called "groundwave" signals, and they are the signals you pick up from your local AM station), and through the air (these are called "skywave" signals, and are most notable at

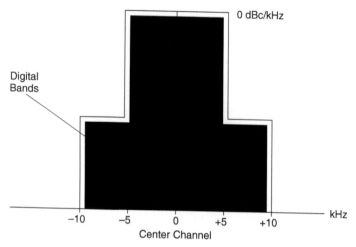

Figure 3.5   Waveform schematic of an all-digital AM-IBOC signal.[29]

night, when they are reflected by the ionosphere to cover great distances). According to system developers, AM-IBOC was developed to accommodate groundwave coverage only.

Figure 3.6 illustrates how AM-IBOC stations on adjacent channels to each other overlap their transmissions.

At night, due to the increased skywave range of many high-power AM stations, the potential for interference expands from a local to a regional or national phenomenon, as demonstrated in Figure 3.7, where interference from previously well-spaced stations may result. Intermodulation of AM-IBOC signals can actually cause interference up to three channels away from the offending station.[30]

A conservatively projective statistical analysis of IBOC interference problems found that approximately two-thirds of AM stations currently licensed to operate at night could most likely do so in a hybrid analog/digital mode without serious problems; most of the remaining third would have to modify their broadcast power and directionalize their signal patterns to avoid or mitigate digitally induced interference. As many as 95 AM stations currently authorized to operate at night might have to cease broadcasting altogether due to unresolvable IBOC interference.[31] Station-to-station AM-IBOC interference manifests itself as a "bacon-frying effect" on analog radio receivers and is not likely to be easily identifiable by listeners as a systemic problem.[32]

The prevailing view of its proponents is that while IBOC has drawbacks regarding the potential for interference between stations, its "improvements" to the radio experience are worth making "compromises" for.[33] Even though many of these interference concerns will go away when all stations cease analog broadcasting, there is no mandated timetable for such a transition, and many in the industry are content to leave these issues unresolved in the meantime.[34]

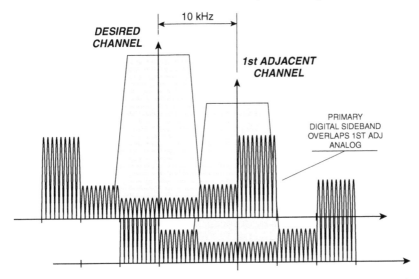

Figure 3.6   Potential for interference between hybrid AM-IBOC stations on adjacent channels.[35]

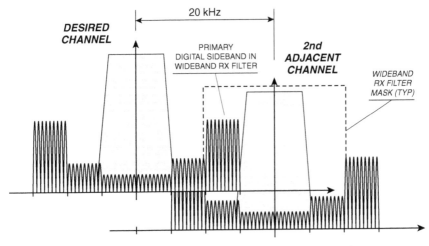

Figure 3.7   Interference between AM-IBOC stations two channels apart—more commonly experienced at night, when AM skywave coverage is greatest.[36]

Consulting engineer Doug Vernier, who worked closely with NPR on IBOC development issues, concludes that "the scale of IBOC interference on analog coverage is unknown," and is most likely to manifest itself in situations where there are "short-spaced stations, overpower grandfathered stations, stations that have contour overlap despite meeting minimum spacing requirements, dual antenna installations, and 'grungy' installations."[37]

Introducing a new digital broadcast service in ignorance of its potential destructiveness to existing analog services, with which it must cohabitate indefinitely, does not seem like wise policy.[38]

The bottom line: IBOC DAB increases the spectral holdings of all incumbent radio stations, while causing greater potential for interference between them. This harms the reception of both analog and digital signals, thereby potentially diminishing the core functionality and reach of radio broadcasting itself.

BANDWIDTH CAPACITY

"Bandwidth" is loosely defined as the capacity to carry digital data. In simple terms, higher bandwidth means increased data capacity and richer, more robust digital media services. As benchmark figures: the average home broadband connection in the United States offers a capacity measured in megabits per second (mbps), with many telecommunications providers offering services measured in the tens or hundreds of megabits. 3G and 4G wireless telephony also offers capacity measured in megabits. U.S. digital television stations can transmit slightly more than 19 mbps of digital data per channel. In contrast, hybrid FM-IBOC stations have a digital bandwidth capacity of approximately 150 *kilobits* per second (in the extended hybrid mode)[39] and AM stations are capable of transmitting just 36 kbps of digital data.[40] Even in their all-digital configurations, FM radio signals can carry just 300 kbps of digital data, while AM tops out at 64 kbps, though these modes have never been substantively tested.[41] Thus, the IBOC DAB system is functionally incapable of providing broadband-quality digital data distribution.

Digital radio's most-touted "improvement" over analog is an increase in audio fidelity, described as "CD-quality" sound for FM and "FM-quality" for AM.[42] True, uncompressed digital audio requires 1.4 mbps of bandwidth, which means IBOC radio signals must significantly reduce the bandwidth of source audio by several orders of magnitude to fit the system's constraints. This is performed by a codec, which stands for "encoder/decoder," and is essentially an algorithm designed to compress digital audio data. Not all codecs are created equal: for example, the widely used MP3 format utilizes an algorithm that reduces the bandwidth of uncompressed digital audio by about 90%.[43] IBOC's proprietors experimented with several codecs before settling on one called Hybrid Digital Coding (HDC).[44] HDC compresses digital audio at a ratio of approximately 15:1 (nearly 95%),[45] using a series of "carefully-engineered perceptual tricks"[46] that attempt to fool the human brain into (hopefully) not perceiving much loss of acoustic range.[47]

Laboratory tests and field observations do not reflect well on the technology's actual audio quality. David Maxson claims that 96 kbps is the "absolute minimum at which a credible [FM-IBOC] stereo signal can be transmitted,"[48] and disputes claims of superior IBOC audio quality: "Compressed digital broadcasts will sacrifice some of the details of musical dynamics that analog FM is capable of delivering."[49] Hybrid FM stations can broadcast a main program channel at 96 kbps, but if they multicast, bandwidth devoted to the main channel is reduced, and any extra program streams must divvy up whatever remaining bandwidth is available. In other words, depending on how a station partitions its digital bandwidth, analog FM may actually sound *better* than IBOC. Since the imposition of digital sidebands preclude an analog AM station from broadcasting in stereo, the perceived relative "improvement" of digital audio quality is accomplished in part by the degradation of the analog transmission.

One engineer involved in IBOC development noted that the system's audio quality is optimized for use in vehicles, where road noise and other distractions can mask many of the codec's compromises.[50] Sony Electronics has commented that the "near-CD" fidelity of IBOC transmissions may "be less tolerable by the public"[51] as the quality of other digital audio distribution technologies improves.[52]

Typically, a mechanical ear is used to measure audio fidelity, but it cannot effectively judge the "perceptual tricks" used in psychoacoustic encoding algorithms.[53] Therefore, IBOC's proponents evaluated the system using a generally selected sample of people, as well as a cadre of "expert listeners" to compare its quality to other source material, most notably analog FM broadcasts.[54] Expert listeners are important because they "are more familiar with peculiar and subtle artifacts" caused by psychoacoustic encoding. In order for such a test to be valid, "The reference [audio sample] must be of the highest quality, and the testing conditions must be designed to attack the codec at its weakest points."[55] Although IBOC's proprietors claim that a perceptual codec performs in such a manner as to be "perceived by typical listeners as 'virtually' the same as a CD,"[56] actual test results do not support this conclusion.

The metric by which the NRSC initially evaluated IBOC audio quality was again one of compromise: not only was its codec's "success" defined simply by the provision of "better than analog" fidelity, but the NRSC's testing criteria contained qualifiers that effectively authorized some level of audio degradation resulting from the intermixing of a station's analog and digital signals.[57] This further lowered the benchmark of audio quality required of IBOC, but the test results of the system's first perceptual audio codec were still unkind.

Among a sample of the general listening population, 82% rated IBOC audio as sounding *equal to* or *worse than* analog FM, as illustrated in Figure 3.8.

Only 18% considered the system's fidelity an improvement. Among expert listeners, shown in Figure 3.9, the results were strikingly similar: 85%

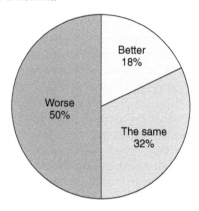

Figure 3.8   General listener grade of initial FM-IBOC audio quality relative to analog FM (n = 1600).[58]

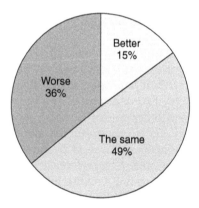

Figure 3.9   Expert listener grade of initial FM-IBOC audio quality relative to analog FM (n = 100).[59]

thought IBOC's audio quality was equal to or worse than analog FM, with just 15% believing that it sounded better.

Tests evaluated by the NRSC in 2004 involving the HDC codec as deployed in today's IBOC system showed some improvement to this state of affairs, but claims of significant improvement over analog broadcasting remain tenuous. The NRSC's evaluative benchmark ultimately declared that FM-IBOC audio quality "should be comparable to or better than the best FM," while AM-IBOC audio quality should "deliver fidelity that approaches present analog FM fidelity."[60] The evaluation showed that FM-IBOC audio quality—when all available bandwidth is devoted to audio transmission—edges out analog FM audio quality in unimpaired situations, but is not qualitatively identical to that of a CD. The results of AM-IBOC testing were more nuanced: although the NRSC concluded that AM-IBOC sounds better than analog AM signals, its quality relative to FM is dependent on what source material is being listened to. For example, listeners found that AM-IBOC was equivalent (or nearly so) to analog FM in samples involving classical and rock music, but speech-heavy programming (including most commercials) sounded worse.[61]

Furthermore, perceptual encoding algorithms do not work well with other forms of digitally encoded audio. When compressed digital audio is re-compressed by another codec, a process called "transcoding" takes place, which can result in the generation of artifacts that further degrades the final output.[62] Many radio stations already use compressed digital audio in some form; encoding it for broadcast through IBOC's algorithm opens up many new possibilities for degraded audio.[63] Therefore, stations are encouraged to rebuild their entire air-chain so that uncompressed audio is fed directly to the IBOC encoder/transmitter, or, if prior encoding is unavoidable, that stations use only one "family" of codecs to minimize transcoding artifacts.[64] This is a hidden cost that must be considered in a radio station's digitalization.[65] IBOC's proprietors have subsequently "clarified" that the "HD" in IBOC's trademarked name, "HD Radio," does not stand for "High Definition,"[66] although early marketing efforts suggested otherwise.[67] This change in the meaning of "HD" has been termed "disingenuous" by *Radio World*.[68]

The paucity of available bandwidth has not deterred IBOC supporters from exploring the system's use beyond the provision of audio, but it is important to note that many of these features involve reducing digital bandwidth devoted to main program audio in order to accommodate them. Several "value-added" features have been developed, such as multicasting (the ability to split an FM-IBOC digital signal into multiple program streams) and datacasting (the provision of digital data for purposes other than audio). Examples of datacasting include iTunes tagging (which allows listeners to "bookmark" songs they hear on the radio for later purchase online), Dolby 5.1 surround sound, traffic and navigation services, recording-on-demand, and subscription audio content.[69] IBOC developers are also pondering several ways to either mimic interactivity or provide such functionality via a separate conduit. Some believe linking digital radio receivers to in-car GPS systems will allow radio stations to serve up locale-specific advertising to listeners. Others propose to link artist, song, and advertiser information conveyed by FM-IBOC stations to enhanced content streamed to mobile phones. There is also a push to explore such backchannel interactivity as a means to provide real-time station audience measurement.[70] Many of these features, such as multicasting, were not originally built into the IBOC feature set: for example, multicasting was primarily developed by NPR following a crash program in 2002. Datacasting is mostly unavailable on AM, where the bandwidth crunch is most extreme.[71] No radio station can provide the complete suite of IBOC's "value-added" features: it may multicast, or broadcast in surround sound, or provide other non-audio services, but the system's limited capacity forces devotion to just a handful of these choices.

Digital radio stations are also not likely to share their spectrum and application windfall with others. Milford K. Smith, Chairman of the NRSC's DAB Subcommittee during IBOC's formative years, has commented that the industry is not interested in accommodating new entrants to radio via the IBOC platform. "Potentially unprofitable demographic segments," he opined, were more likely to find the Internet a better place "to seek mass distribution of their product."[72]

The bottom line: digital radio operates in an environment bereft of bandwidth. This forces a complicated series of compromises to IBOC's audio quality and ability to provide additional services. Assertions that IBOC represents a significant qualitative improvement over the legacy analog broadcasting service rest on shaky ground.

PROPRIETARY NATURE

Perhaps the least-discussed but most important detriment of IBOC digital radio is its wholly proprietary nature. After a few years of developmental competition, IBOC's two major proprietors (Lucent Digital Radio and USA Digital Radio) merged to form iBiquity Digital Corporation in 2000 (see Chapter 4). With the creation of iBiquity, all of the intellectual property used in IBOC technology was consolidated under one roof.[73] iBiquity has no qualms with leveraging this position; in fact, it is inherent to its business model.

For broadcasters, there is a one-time $12,500 licensing fee payable to iBiquity in order to implement the technology—this fee was initially $25,000 per station, but has been halved to entice broadcaster adoption.[74] For a time, iBiquity considered setting the license fee based on a station's market size and audience share, but ultimately dismissed that formula as overly complicated.[75] If an FM station chooses to multicast, it must pay iBiquity an amount equal to 3% of quarterly net revenue or $1,000 per year, whichever is greater, for each new digital channel. Utilizing IBOC's datacast functionality also requires revenue-sharing with iBiquity.[76] The company will not charge for software upgrades that fix bugs in the IBOC system, but as new features are added to the protocol those, too, may come at a price.[77] For example, iBiquity has acquired the intellectual property rights to technology that would allow the encryption of subscription-based programming on the IBOC platform, so fees to use that feature will also flow to the company.[78]

iBiquity's broadcaster licensing terms clearly state that the contract is perpetual; that a broadcaster does not own the IBOC software; and it may not sell or otherwise transfer IBOC technology to another party, except in cases where a radio station is sold. Such a transfer may only occur after written permission is obtained from iBiquity.[79] Furthermore, the company retains the right to audit a radio station's financial records at any time and revoke a station's license agreement for nonpayment.[80] If a station finds itself in fiscal difficulty, iBiquity may allow the broadcaster to continue to use its system, provided a 1.5% interest charge is tacked onto the station's unpaid license balance.[81] If the IBOC system does any material damage to a radio station, the company is indemnified for costs in excess of $200,000.[82] These terms were promulgated against the recommendations of iBiquity's own broadcaster advisory council.[83]

As a result, iBiquity Digital Corporation wholly controls who may broadcast in the U.S. digital radio domain. The net effect is that broadcasters now require two licenses to operate digitally—one from the FCC to utilize an AM or FM channel, and one from iBiquity to broadcast in a digital mode. Until now, such gatekeeper authority rested solely in the hands of the FCC, which seems unconcerned about this important power shift.

For digital radio receiver manufacturers, the terms of license are similarly strict. Licenses are nontransferable, all products that contain IBOC technology must be labeled on the outside as such, derivative works are prohibited under penalty of law, and license terms are only good for five years. Receiver manufacturers must also pay a one-time fee to iBiquity (the amount of which is undisclosed) and a "per unit" royalty "based on a percentage of the aggregate total gross invoiced [receiver] sales." Per-unit royalties are paid quarterly. Receiver manufacturers must also agree to open their books to iBiquity and submit "quarterly sales reports along with royalty payments."[84] Adding IBOC functionality is expensive relative to other receiver features: whereas a chipset to provide analog FM radio reception might cost a consumer electronics manufacturer a few cents, with no perpetual charge, the price of an IBOC receiver chipset is in the range of several dollars and comes with residual costs.[85]

Finally, any developer of IBOC-related applications must first run their proposals by iBiquity and, if they are approved, must pay to use the programming language that the company has written for application development.[86] The FCC's operative governance on this intellectual property issue dates back to a rulemaking from 1961—well before the advent of modern computing, not to mention the intricate issues that come with the constraints of proprietary technology in the realm of software.[87]

Initially, public radio stations strongly opposed IBOC's proprietary nature. At the 2002 Public Radio Conference, Nevada Public Radio General Manager Lamar Marchese asked an FCC representative, "How is the FCC coming down on this perpetual franchise granted with no competition?" The question received much applause, but no substantive answer. Instead, iBiquity President and CEO Robert Struble responded nonchalantly, "We are a business. We do have to make some money."[88] Public broadcasters subsequently banded together to seek some relief from iBiquity's pay-to-play system:[89] iBiquity and the Corporation for Public Broadcasting hammered out a deal for CPB-qualified stations that waived the one-time license fee for early adopters, capped it at a discounted rate for the rest, and removed the perpetual fees for the use of applications such as multicasting and datacasting.[90]

Perhaps most importantly, technical information about the IBOC standard has not been fully disclosed. According to the terms of the National Radio Systems Committee's broadcast standard-setting process, all standards must include two forms of documentation: normative and informative. Normative documents contain "a detailed description of a component of the standard. To be compliant with the standard, a device or system *must* satisfy not only the general criteria in the main standard document, but also the detailed criteria contained in those normative references that relate to the workings of the device." All normative documents included in the NRSC's IBOC standard remain the property of iBiquity Digital Corporation.[91] Informative references, on the other hand, are used to "help the reader understand something about the standard that may not be evident by reviewing the . . . normative references. An informative reference may contain, for instance, an example of an implementation of the standard, or material that provides background information on a specification." iBiquity has declined to provide complete normative or informative references for the codec used in its system, leaving those with an interest in exploring IBOC without a complete set of documentation with which to tinker.[92] The company explained that it had "compelling reasons" for withholding this information, and its recalcitrance on the issue forced the NRSC to publish an IBOC standard that does not technically include a codec. This provides iBiquity with the opportunity to change or modify this aspect of its system with no meaningful review or oversight, which could add future costs to system implementation.[93]

iBiquity's refusal to disclose this information also forces transmitter and receiver manufacturers into the company's intellectual property straitjacket. According to David Maxson, "In theory, a manufacturer could develop a product that is . . . compliant but that has not earned the right to use the HD Radio name and logo. In practice, the manufacturer would find it necessary to license iBiquity patents to manufacture and sell its . . . product. While doing so, the manufacturer may be enticed . . . to take the next step and join the HD Radio family."[94] Nondisclosure also subverts patent limitations on IBOC technology.

Although the National Radio Systems Committee ostensibly approves standards that are open on a Reasonable and Nondiscriminatory (RAND) basis,[95] it is difficult to see how that applies to the IBOC digital radio system. In *Radio World*, Skip Pizzi lamented that "there is something fundamentally troubling about a digital broadcasting format designed by a unilateral, proprietary group and not driven by an open standards process . . . If the FCC rubber-stamps a format proposed by private interests, it will be neglecting due diligence and abdicating its ultimate responsibility as a steward of the public interest."[96] Pizzi was not alone in this criticism: Paul Signorelli, chief technology officer for Impulse Radio, a developer of digital radio broadcast applications, opined that a proprietary approach would inhibit innovation in the IBOC arena. "Ironically, it seems iBiquity would rather own all of nothing rather than a large piece of something huge," he wrote. "So let's not simply let them assuage us with talk of openness. We're dumb, but not that dumb."[97] David Maxson agreed:

> With such dominance of the technology comes the potential for too much control in the hands of one enterprise. As the regulatory authority, the FCC should be certain that all standards and policies encourage competition in all levels of the IBOC marketplace . . . Innovators should not be forced to get the permission of the dominant competitor to develop new ideas. Licensure of the core technologies should be at arms length from the activities that develop features that utilize the technology.[98]

Suspicion of IBOC's proprietary nature can be found among all constituencies involved in the digital radio issue, but is most severe among independent broadcasters, consulting engineers, and the public. Many worry that by adopting this technology, the United States is thwarting global digital radio compatibility, creating a universe of "undesirable diversity" in an embryonic technological space.[99] For its part, iBiquity has demonstrated a powerful sense of hubris when confronted with these concerns. At the 2002 NAB annual convention, Robert Struble directly likened the company to Microsoft: "We're a software company. If you buy a transmitter, you'll need new software from us."[100] That same year, iBiquity claimed it would reap some $600 million per year in licensing revenue by 2010.[101]

In addition to the license fees, there really is no "average" cost associated with adding IBOC functionality to a radio station; the price of the "upgrade" will vary depending on the age and existing condition of its transmitter plant and the need to revamp a station's air-chain to purify the digital audio stream and remove points where transcoding takes place.[102] In the long run, generally speaking, larger stations will pay more to broadcast in IBOC than smaller stations.[103]

The bottom line: the proprietary nature of IBOC adds an unprecedented cost to the business of broadcasting and fundamentally shifts control of who may access the public airwaves from a purely public gatekeeper (the FCC) to what is effectively a public–private partnership (the FCC and iBiquity). This cost, coupled with iBiquity's lack of transparency regarding the core functionality and development of its technology, dampens interest in digital radio and discourages innovation with the protocol.

The three fundamental flaws of IBOC digital radio cannot be overcome through redesign or other remedial measures. The spectral footprint of IBOC signals may cause destructive interference to both host broadcasters and neighboring stations on the dial; in the case of AM-IBOC, the potential impact of interference may be regional or national in scope. With regard to bandwidth capacity, IBOC provides very little compared to other twenty-first-century digital information conduits. It falls far short of providing broadband-level service, and its lack of interactivity is a hindrance with which IBOC developers and broadcasters continue to struggle. The promise of providing radio service that sounds "better than analog" is a dubious one; furthermore, IBOC's "value-added services," such as multicasting and datacasting, are underwhelming and deployed at the sacrifice of digital audio fidelity. Finally, the end-to-end proprietary nature of IBOC stifles its uptake and ongoing development. Ironically, when the technology's supporters formally engaged the FCC in policymaking to cement the standard as the future of U.S. digital radio, these issues were hardly on the agency's radar. The result was a controversial standards-setting process in which IBOC's deficiencies would repeatedly come back to haunt it.

NOTES

1. NRSC DAB Subcommittee, "Evaluation of USA Digital Radio's Submission to the NRSC DAB Subcommittee of Selected Laboratory and Field Test Results for its FM and AM Band IBOC System," April 8, 2000, filed as *ex parte* notice by the National Association of Broadcasters, MM 99–325, May 12, 2000, 5.
2. David P. Maxson, *The IBOC Handbook: Understanding HD Radio Technology* (Burlington, MA: Focal Press, 2007), 80.
3. Ken C. Pohlmann, *Principles of Digital Audio, 5th Edition* (New York: McGraw-Hill, 2005), 653.
4. *Ex parte* Letter from iBiquity Digital Corporation, MM 99–325, November 1, 2000, 5.
5. See 47 CFR Sec. 73.317 (FM) and 47 CFR Sec. 73.44 (AM).
6. Comments of Scott A. Clifton, MM 99–325, May 25, 2004, 1–2.
7. Federal Communications Commission, *Notice of Proposed Rulemaking: Digital Audio Broadcasting Systems and Their Impact On The Terrestrial Radio Broadcast Service*, Audio Services Division, Mass Media Bureau (MM) Docket 99–325, November 1, 1999, 4–5.
8. Maxson, 241, 256.
9. Pohlmann, 654, 656.
10. *Ex parte* Letter from iBiquity Digital Corporation, MM 99–325, November 1, 2000, 7.
11. See Mike Starling, "Call It IBAC," *Radio World*, May 22, 2002, 62; Robert Gonsett, "Call It IBAC," *Radio World*, May 22, 2002, 62; Comments of Barry McLarnon, MM 99–325, June 14, 2004, 12; and Comments of Clifton, 1.
12. Maxson, 246–247, 255; and USA Digital Radio, Inc., *Submission of Report*, MM 99–325, December 21, 1999, Appendix F, 6, 20.
13. See Steve Behrens, "More Power for HD Radio, More Buzz on Analog," *Current XXVII*, no. 15 (September 2, 2008), 8.
14. Barry McLarnon, "Skip Pizzi and HD," *Radio World*, June 6, 2007, 53.
15. NAB and CEA (on Behalf of the NRSC), *Evaluation of the iBiquity Digital Corporation IBOC System Part 1—FM IBOC*, MM 99–325, December 3, 2001, 14.
16. See NPR Labs, *Report to the Corporation for Public Broadcasting: Digital Radio Coverage & Interference Analysis (DRCIA) Research Project, Final Report*, May 16, 2008, 21, and Maxson, p. 52, 421.
17. Russ Mundschenk and Milford Smith, "Why a Full 10 dB Increase Is Necessary," *Radio World Engineering Extra*, October 14, 2009, 16.

18. NPR Labs, 42–43.
19. Ibid., 31.
20. Tom Vernon, "Ask the Experts? OK, Let's Do," *Radio World*, September 8, 2010, p. 30, 32.
21. See USA Digital Radio, Inc., *Submission of Report*, MM 99–325, December 21, 1999, 9–10, and Ibid., Appendix B, 20.
22. Ibid., Appendix H, 17.
23. National Association of Broadcasters et al., *Report of the NRSC Evaluation of AM IBOC*, MM 99–325, April 16, 2002, 15.
24. Maxson, 351.
25. Comments of Lucent Digital Radio, Inc., MM 99–325, January 24, 2000, Appendix I, 3.
26. Maxson, 81. In 2010, iBiquity unveiled an alternate waveform schematic that eliminates the sharing of spectrum by analog and digital signals; this improves the analog audio quality but changes nothing regarding the hybrid signal's spectral footprint. See John Anderson, "AM-HD Undergoes Radical Redesign," *DIYmedia.net*, April 3, 2010, http://diymedia.net/2010/04/03/am-hd-undergoes-radical-redesign/.
27. See Jeff R. Detweiler, "Conversion Requirements for AM & FM IBOC Transmission," iBiquity Digital Corporation White Paper, n.d., http://www.ibiquity.com/i/pdfs/Conversion_Requirements.pdf, 4; Comments of USA Digital Radio, Inc., Appendix I, 4, footnote 2; and Pohlmann, 658.
28. Steven A. Johnson, "The Structure and Generation of Robust Waveforms for AM In Band On Channel Digital Broadcasting," iBiquity Digital Corporation White Paper, n.d., http://www.ibiquity.com/i/pdfs/Waveforms_AM.pdf, 4.
29. *Ex parte* Letter from iBiquity Digital Corporation, MM 99–325, November 1, 2000, 8.
30. Maxson, 258.
31. John Anderson, "Digital Radio in the United States: Privatization of the Public Airwaves?" *Southern Review* 39, no. 2 (2006), 10.
32. Michael LeClair, "From the Tech Editor: The AM IBOC Digital Conundrum," *Radio World Engineering Extra*, February 22, 2006, 3.
33. NAB and CEA (on Behalf of the NRSC), Submission of Report, NRSC DAB Subcommittee, *Evaluation of The iBiquity Digital Corporation IBOC System, Part 1—FM IBOC*, November 29, 2001, MM 99–325, December 3, 2001, 24.
34. Paul J. McLane, "From the Editor: Will Radio Ever Be All-Digital?" *Radio World*, November 25, 1998, 4.
35. National Association of Broadcasters et al., 16.
36. National Association of Broadcasters et al., 17.
37. Quoted in Daniel Mansergh, "HD Radio: 'Pieces Are Just Loose,'" *Radio World*, January 3, 2007, 10.
38. Steve Davis, "Can AM Analog and Digital Co-Exist?" *Radio World*, June 7, 2006, 24.
39. See Steve Behrens, "Field Testing Resumes for Radio's Digital Best Hope," *Current* XVIII, no. 15 (August 16, 1999), 19; Mike Janssen, "With Second Channel, FM Branches Out," *Current* XXIII, no. 8 (May 10, 2004), 1; Maxson, 6; and USA Digital Radio, Inc., *Submission of Report*, Appendix B, 4.
40. See iBiquity Digital Corporation: "Market Facts," January 2007, http://www.ibiquity.com/i/january_%202007.pdf, 4, and Maxson, 91.
41. Comments of Lucent Digital Radio, Inc., 7.
42. See HD Digital Radio Alliance, "What Is HD Radio," http://www.hdradio.com/what-is-hd-radio.
43. Thomas Mock, "It's All About the Algorithm—But Which One Will Win?" *IEEE Spectrum*, September 2004, 46.

44. Pohlmann, 651.
45. Maxson, 93.
46. Reply Comments of David Maxson, MM 99–325, February 23, 2000, 2.
47. Pohlmann, 315, 330.
48. Reply Comments of Maxson, 19.
49. Ibid., 3.
50. Detweiler, "Conversion Requirements for AM & FM IBOC Transmission," 3.
51. Comments of Sony Electronics Inc., MM 99–325, January 24, 2000, 4.
52. Ibid., 5.
53. Pohlmann, 404.
54. See Ronald K. Jurgen, "Broadcasting With Digital Audio," *IEEE Spectrum*, March 1996, 56.
55. Pohlmann, 411.
56. Comments of USA Digital Radio, Inc., Appendix B, 5–6.
57. Maxson, *The IBOC Handbook*, 45.
58. Data compiled from Comments of Lucent Digital Radio, Inc., Appendix G, 10, fig. 1a.
59. Data compiled from Ibid., Appendix G, fig. 1b.
60. National Radio Systems Committee, *Evaluation of iBiquity AM and FM IBOC "Gen 3" Hardware*, NRSC R-206, June 30, 2004, 2. This report provides a summary of conclusions gleaned from the listening tests; the actual protocol and data of the testing regimen itself have never been publicly released.
61. Ibid., 3–4.
62. Pohlmann, 347–348.
63. Comments of Kevin M. Tekel, MM 99–325, June 18, 2002, 1.
64. Detweiler, "Conversion Requirements for AM & FM IBOC Transmission," 2.
65. See iBiquity Digital Corporation, "Quality Audio Processing," n.d., http://www.ibiquity.com/broadcasters/quality_implementation/quality_audio_processing; Detweiler, "Conversion Requirements for AM & FM IBOC Transmission," 3; Detweiler, "Digital STLs," *Radio World*, December 5, 2001, 8; Editorial, "Listening With Analog Ears," *Radio World*, May 5, 2004, 46; and Maxson, 406–409.
66. See HD Digital Radio Alliance, "Talking About HD Radio Technology," n.d., http://hdradioalliance.com/talking_about_hd_radio_technology; "Promoting HD Technology On Your Site," n.d., http://hdradioalliance.com/promoting_hd_radio_technology_on_your_site; and iBiquity Digital Corporation, "Trademarks," n.d., http://www.ibiquity.com/about_us/trademarks. iBiquity later suggested that "HD" stood for "Hybrid Digital," and now considers it to mean nothing at all.
67. See Clear Channel Creative Services Group/HD Digital Radio Alliance, "Are You Def Yet?" http://web.archive.org/web/20120127083124/http://www.areyoudefyet.com/; John Anderson, "Wack Spots Promote HD Radio," *DIYmedia.net*, April 22, 2006, http://diymedia.net/2006/04/22/wack-spots-promote-hd-radio/; and Paul McLane, "From the Editor: Radio Finally Starts to Go Def," *Radio World*, May 10, 2006, 4.
68. Paul McLane, "From the Editor: Why Ibiquity is now iBiquity Again," *Radio World*, October 22, 2008, 4.
69. See Michael Starling, "Digital Radio *Could* Be More Satisfying, But Will It?" *Current* XXI, no. 6 (March 25, 2002), 17, 21; Mike Janssen, "IBOC Developer Discusses Royalties, Secondary Audio at PRC," *Current* XXI, no. 10 (June 3, 2002), 3; HD Digital Radio Alliance, "iTunes Tagging," n.d., http://www.hdradio.com/what-is-hd-radio#features-iTunesTagging; iBiquity Digital Corporation, "New Mobile Service Providers," n.d., http://www.ibiquity.com/

automotive/new_mobile_service_providers; and iBiquity Digital Corporation, "HD Radio: Experience It," n.d., http://www.ibiquity.com/hd_radio/hdradio_experience_it.

70. Anderson, 13.
71. Skip Pizzi, "Radio Copes With Insufficient Data," *Radio World*, July 14, 2004, 27.
72. Quoted in Michael P. McCauley, "Radio's Digital Future: Preserving the Public Interest in the Age of New Media," in Michele Hilmes and Jason Loviglio, eds, *Radio Reader: Essays in the Cultural History of Radio* (New York: Routledge, 2002), 521.
73. Maxson, 399.
74. See iBiquity Digital Corporation, "HD Radio Broadcaster Licensing Fact Sheet," March 2013, http://www.ibiquity.com/i/Radio%20Broadcaster%20Licensing%20Fact%20Sheet%203-2013.pdf; iBiquity Digital Corporation, "HD Radio Broadcaster Licensing Fact Sheet," January 2009, http://www.ibiquity.com/i/Licensing_%20Fact_%20Sheet_2009.pdf; and iBiquity Digital Corporation, "2012 Station License Agreement (for stations eligible for discount pricing)," 2012, http://www.ibiquity.com/i/2012%20Form%20SLA-A%20STANDARD.pdf, 1–12.
75. Skip Pizzi, "IBOC Raises Eyebrows," *Radio World*, March 27, 2002, 67.
76. See iBiquity Digital Corporation, "HD Radio Broadcaster Licensing Fact Sheet (2013)"; and iBiquity Digital Corporation, "2012 Station License Agreement," 4.
77. See iBiquity Digital Corporation, "HD Radio Broadcaster Licensing Fact Sheet (2013)"; iBiquity Digital Corporation, "2012 Station License Agreement," 5; and Leslie Stimson, "iBiquity Gets Specific With Fees," *Radio World*, March 16, 2005, 1, 16.
78. Leslie Stimson, "iBiquity to License Radio Content-on-Demand," *Radio World*, September 1, 2002, 5.
79. See iBiquity Digital Corporation, "HD Radio Broadcaster Licensing Fact Sheet (2013)"; and iBiquity Digital Corporation, "2012 Station License Agreement."
80. iBiquity Digital Corporation, "2012 Station License Agreement," 4, 5, 7.
81. Ibid., 5.
82. Ibid., 6.
83. Leslie Stimson, "IBOC Fees Stir Reaction," *Radio World*, April 10, 2002, 10.
84. iBiquity Digital Corporation, "Becoming an HD Radio IP Licensee," n.d., http://www.ibiquity.com/manufacturers/receiver_manufacturers/license_agreement.
85. Leslie Stimson, "And Beyond Zune?" *Radio World*, July 1, 2009, 5.
86. Paul Signorelli, " 'Open' Source? Up for Interpretation," *Radio World*, September 10, 2003, 15.
87. See *Revised Patent Procedures of the Federal Communications Commission*, Public Notice (Dec. 1961), 3 FCC 2d 26 (1961).
88. Mike Janssen, "IBOC Developer Discusses Royalties, Secondary Audio at PRC," *Current* XXI, no. 10 (June 3, 2002), 3.
89. Skip Pizzi, "IBOC DAB, in the Public Eye," *Radio World*, June 19, 2002, 14, 22.
90. See Mike Janssen, "Sound Upgrade Restarts Radio's Move to Digital," *Current* XXII, no. 15 (August 25, 2003), 16; Mike Janssen, "Two-channel Digital FM 'Works Great'," *Current* XXIII, no. 1 (January 19, 2004), 11; "Quick Takes: CPB Deal with iBiquity Could Save Stations Thousands," *Current* XXIV, no. 12 (June 27, 2005), 4; and "Quick Takes: CPB Awards Digital Radio Grants, Buys Group License From iBiquity," *Current* XXIV, no. 19 (October 17, 2005), 4.

91. Maxson, 66.
92. Ibid., 68, 79.
93. See Leslie Stimson, "IBOC Standards Are On, Again," *Radio World*, March 28, 2004, 8, and Leslie Stimson, "Digital Radio Vote Is Yes: NRSC OK's IBOC Standard," *Radio World*, May 13, 2005, 5.
94. Maxson, 68.
95. Ibid., 66.
96. Quoted in Anderson, 15.
97. Signorelli, 15.
98. Comments of David Maxson, MM 99–325, June 17, 2004, 3.
99. Skip Pizzi, "Whups, There Goes the Neighborhood," *Radio World*, February 11, 2004, 14.
100. Quoted in Stimson, "IBOC Fees Stir Reaction," 10.
101. Anderson, 15.
102. Bob Clinton, "How to Prep Now for IBOC Later," *Radio World*, May 23, 2001, 8, 20.
103. Richard J. Fry, "FM IBOC Planning Worksheet," *Radio World*, February 13, 2002, 20.

4 FCC "Deliberation" of HD Radio

Although the question of whether analog and digital radio signals could peacefully coexist remained unresolved, USA Digital Radio forced the issue by filing a Petition for Rulemaking with the Federal Communications Commission on October 7, 1998, asking it to "immediately make a finding that IBOC will be the method of transmission for DAB in the United States."[1] USADR asserted that the adoption of any other technology would "create tremendous turmoil in the radio industry, disrupting service to the public, and impose a significant administrative burden on regulatory authorities."[2] It admitted that "a number of technical tradeoffs" would be required to make IBOC work, but downplayed their implications.[3] USA Digital Radio also claimed its technology had wide support within the radio industry, touting equity investment from 12 of the largest radio broadcasters in the United States, as well as an infusion of funding from Chase Capital Partners.[4]

In reality, a minority of station-owners supported the IBOC concept, though they happened to control the majority of industry revenue. This, combined with a qualified endorsement from National Public Radio, engendered a perception among regulators that consensus existed within the industry to adopt the technology. From a neoliberal policy perspective, an economic consensus was all that was necessary to sign off on IBOC's proliferation; the FCC took many of its proponents' assertions about the technology's functionality and features at face value. Dissenting voices, of which there were many, were effectively placed outside the realm of acceptable policy discourse defined by the Petition for Rulemaking itself.

The FCC opened a 75-day public comment window on the petition a month after it was filed.[5] Many large national broadcasters, such as Disney, CBS, and the Gannett Company (the latter two being founders of USADR) agreed that expedited approval of the technology was critical.[6] Cumulus Media succinctly summed up the primary argument of proponents: "The driving force behind this digital revolution is clear; for most applications, digital is simply superior to analog."[7] Other broadcasters urged the FCC to take greater care with the evaluation and eventual approval of any digital radio technology. Greater Media and the Radio Operators Caucus (an "informal group" of 20 radio broadcast engineers employed with regional broadcast conglomerates that represented "hundreds of . . . stations") agreed in principle with a digital radio transition but did not want USADR's technology to be chosen as the de facto standard simply on the basis that it was the first to file a Petition for Rulemaking.[8] National Public Radio cautioned the FCC not to foreclose consideration of "alternative approaches to DAB if IBOC proves infeasible or so burdened with compromises that the benefits of a digital transition are fundamentally undermined."[9] Clear Channel practically begged the FCC to referee further development of the technology in order to avoid implementation hurdles such as those that plagued the digital television transition process.[10]

Independent broadcasters and consumer electronics manufacturers expressed significant concerns about the fundamental deficiencies of IBOC. Big City Radio, a company with a handful of stations on the peripheries of major markets, asserted that the technology "risks significant—and even fatal—interference to many existing AM and FM licensees."[11] It also highlighted the economic momentum behind the petition: "The Commission should not credit the advocacy of USADR's investors as a reason for moving hastily on its Petition, especially when USADR's proposed technology has been only incompletely simulated and even less sufficiently tested."[12] Ford Motor Company told the FCC that it could only support an IBOC system so long as it would not harm existing analog signals—a guarantee the technology could not make.[13] Similarly, the Consumer Electronics Association emphasized the need for a complete test-record of IBOC performance.[14]

Members of the public were highly skeptical of USADR's petition. The Citizen's Media Corps of Brookline, Massachusetts noted that the industry had initially supported an alternate-band DAB system, and it still believed that such a system "would be in the best interests of the citizens of the United States . . . It is by no means a foregone conclusion that new spectrum cannot be found. What is not lacking is spectrum, it is the will to prioritize finding that spectrum over the objections of the very powerful radio industry lobby."[15] The Prometheus Radio Project, a primary facilitator of the development of new, noncommercial low-power FM (LPFM) radio stations,[16] recommended "the FCC find independent data about what the American public really wants from radio before going ahead with a plan that ignores the potentialities for more channels."[17]

USA Digital Radio noted these concerns but asserted that its technology was "in a final stage of development and there is consensus on an IBOC approach. Hesitation on the part of the Commission is unnecessary and will prolong the wait for digital radio." To amplify the perception of support within the radio industry, USADR referenced "a recent study of broadcasters" that claimed, "approximately 30% of radio stations are likely to convert to digital broadcasting within the first two years, with 56% likely to convert within the first five years."[18] Meanwhile, IBOC's proponents launched a personal lobbying campaign at FCC headquarters: USADR representatives met with FCC Commissioners, their advisors, and senior staff nearly 20 times in 1999 to advance the arguments for IBOC's adoption and press for a quick decision cycle.[19]

The actual developmental status of IBOC remained muddled, according to industry perspectives documented in the trade press. In an interview with *Current*, the newspaper of record for public broadcasting, NPR's chief technology officer Don Lockett conceded that USADR "has the lead in clout, if industry politics count for anything at the FCC." However, he was also surprisingly candid about the potential for IBOC to fail. "This is the last round for IBOC, I would say," he told *Current*. "Based on what I learned in the first round of testing, it has to work [soon] or we have to move on to something else."[20]

FRAMING THE POLICY DEBATE

On November 1, 1999, the FCC issued a Notice of Proposed Rulemaking (NPRM) to formally explore the concept of terrestrial digital radio.[21] Regulators contextualized the effort as a means "to significantly enhance the American radio broadcast service."[22] The FCC acknowledged that IBOC had "not been conclusively proven to be technically viable at this point in time, yet great strides have been made and the systems certainly hold real promise." Even though "other documentation" demonstrated that alternate-band digital radio technology might be superior to IBOC, "No proponent of a Eureka-147 or other non-IBOC DAB system has filed comments in response to USADR's *Petition*. We currently are unaware of any such proponents in the United States." This was not true: independent broadcasters, consumer electronics manufacturers, and the public all suggested the exploration of non-IBOC alternatives. In short, the FCC was predisposed to declare other digital radio systems infeasible, as the domestic radio industry had no pecuniary interest in them.[23] The FCC also dismissed the idea of establishing an advisory committee to study digital radio, citing its historical use of the NRSC for the advancement of new radio technologies, and concluded, "We believe that it is necessary and appropriate to rely to some degree on the expertise of the private sector for DAB system evaluations and, ultimately, recommendations for a transmission standard."[24]

This left industry proponents of IBOC with three nominally competitive systems to consider and the task of consolidating them into a single digital radio standard for FCC certification. On December 21, 1999, USA Digital Radio filed notice with the FCC detailing the testing it had done of its IBOC system.[25] The report also revealed that USADR and Digital Radio Express had effectively merged their development efforts; USADR would continue to tweak IBOC technology, while DRE would concentrate on the development of applications for it.[26] DRE's capitulation in the race to own the U.S. digital radio standard further positioned USADR as the front-runner over Lucent, which filed its own technical synopsis with the FCC detailing its

IBOC development program.[27] Lucent did candidly admit that there would be tradeoffs between analog and digital reception quality on the AM and FM bands, but such compromises were "variables" that should be left up to individual stations to accommodate.[28]

Shortly after USADR's technical filing, the FCC was spiked with dozens of boilerplate comments from supposedly independent broadcasters endorsing the USADR system and asking for its immediate promulgation.[29] They praised the IBOC protocol and vilified any notion of an alternate-spectrum digital radio option. Several USADR investors, in conjunction with the NAB, coordinated the filing of these complementary IBOC assessments.[30] USADR urged regulatory approval of its technology by the end of 2000, "in order to avoid a stalemate between different components of the industry."[31] By the close of 1999, USADR claimed the financial backing of radio conglomerates that boasted "coverage in 196 of the 270 Arbitron-rated markets, access to 200 million listeners and combined revenues equating to 46% of the radio industry's total revenues."[32] More than "forty world-class engineers and scientists" were tasked to its IBOC project, which already had "27 patents, with numerous pending patent applications covering broad aspects of [the] technology."[33]

At the National Radio Systems Committee, however, USADR was reluctant to open IBOC up to meaningful examination. The company declared it would provide "only the level of information it views as necessary to demonstrate an improvement over analog." The withholding of data was justified by a desire to protect proprietary information.[34] Even so, the NRSC told the FCC that it still could "develop a testing process and measurement criteria that will produce conclusive, believable and acceptable results, and be of a streamlined nature so as not to impede rapid development of this new technology."[35] The committee did admit that "a number of compromises and tradeoffs among key aspects of the system" would have to be made in order to make IBOC function in the real world, and that assessing whether IBOC represented "a significant improvement over analog services" would be "challenging."[36] But with control of the NRSC now firmly in the hands of IBOC proponents, the committee's ultimate endorsement of the technology seemed all but guaranteed.

In the wake of the FCC's NPRM, several important constituencies qualified their advocacy of IBOC. National Public Radio cautioned that the commission's operational goals for such systems were "simply unrealistic": the technology could not provide increased signal robustness, improved audio fidelity, and new digital program or data services all at the same time.[37] It urged the commission to stay proactively engaged in the testing and evaluation of digital radio systems.[38] Consumer electronics manufacturers again attempted to emphasize IBOC's fundamental deficiencies. Sony, the world's largest manufacturer of consumer electronics, asked the FCC to select a standard that could effectively compete against other digital audio distribution systems, which were multiplying quickly;[39] it also suggested IBOC was simply not robust or flexible enough to entice listener interest.[40] Visteon Automotive Systems, one of the world's largest producers of standard-equipment radio receivers, observed that if IBOC's drawbacks manifested themselves in the real world, they would cause consumer confusion and dismay.[41] Visteon even proposed a public education program to explain to listeners how and why their existing analog radio reception might worsen in the wake of IBOC's adoption.[42]

The Consumer Electronics Association (CEA) boiled down its critique of IBOC to a finer point, challenging the technology's proponents to "convincingly demonstrate that an IBOC-delivered audio experience is attractive to listeners [and] sufficient to persuade consumers to purchase new radio receivers."[43] With more than 700 million analog radio receivers in use in the United States, CEA implored the commission to become better versed in the negative impacts of IBOC technology on existing analog radio service, and urged the FCC to keep open the possibility of adopting another DAB technology.[44] To that end, CEA submitted a 42-page concept analysis to the FCC of a "new mobile multimedia broadcast service" deployed on reclaimed swaths of analog UHF television spectrum.[45] This would be the last time that any major industry-related constituency would substantively raise the notion of alternatives to IBOC in policy discourse.

Consulting engineer David Maxson, a member of the NRSC's DAB Subcommittee, was well aware of IBOC's technical deficiencies. "Because of the limited data capability," Maxson told the FCC:

> IBOC in its hybrid mode (with analog still present) does not appear to break radio broadcasting into this brave new world in the manner that DTV appears to do for television ... The main feature of a hybrid IBOC system is its anticipated improved robustness over analog. Is improved robustness enough of a draw to get people to change out millions of analog radios for new DAB radios over a few years?

"If radio broadcasters did not fear that their investments would be at risk for proposing a DTV-like solution, we suspect one might have been developed," he observed.[46] Ultimately, Maxson challenged the FCC to examine radio's digital transition from the perspective of the radio audience rather than on the basis of economic rationales proffered by the industry. "The Commission has a duty to be the advocate for consumer benefit in selecting a final approach. Because the nature of free over-the-air broadcasting requires ubiquitous, low cost reception, it should not be left to the marketplace to decide how it should be done."[47]

Independent broadcasters and the public were openly skeptical of IBOC. Gene A. Benedictson, the owner of small stations in Washington state, noted, "So far, all of the money and support seems to be coming from the huge broadcasting groups that have huge money resources to dip into. Small market radio can't afford to spend money like that."[48] Several individuals were mystified that a regulatory agency specializing in communications technology might overlook the potentially detrimental nature of expanding the spectral footprint of all radio stations, thereby exacerbating the potential for interference between them.[49] The Virginia Center for the Public Press provided the first independent report of FM-IBOC interference, monitored on two separate receivers during USADR field tests.[50] Others questioned the notion that any digital radio system was a de facto improvement over analog broadcasting.[51] Several radio listeners suggested the FCC leave existing analog radio service alone and implement digital broadcasting on other spectrum.[52] Ted M. Coopman of Santa Cruz, California, writing on behalf of 15 community media groups and 29 individuals from 13 states, noted that an alternate-band digital radio system could raise billions of dollars in revenue from the auction of new frequencies: "Simply granting additional spectrum to broadcasters . . . for this conversion is nothing more than a give-away of national resources to for-profit business interests."[53]

Proponents of IBOC ignored, dismissed, or misconstrued the concerns of critics. USA Digital Radio asserted "the comments of a few small broadcasters expressing concern about the cost of IBOC reflect a misunderstanding of the nature of the flexible IBOC implementation process rather than a fundamental disagreement about a transition to DAB."[54] Lucent Digital Radio claimed that "there is no justification for AM and FM radio broadcasters to be left in the analog world, notwithstanding the Commission's decision on whether additional spectrum is justified for radio broadcasting."[55] Both Lucent and the NAB chastised the Consumer Electronics Association for casting doubt on IBOC's viability by questioning the integrity of CEA's analysis.[56]

In a pernicious use of the FCC's *ex parte* rules, the National Association of Broadcasters submitted the National Radio Systems Committee's first digital radio system evaluations on May 12, 2000—more than three months after the window for initial public comment on the proceeding had closed. The NRSC followed the wishes of IBOC proponents and limited its technical review to those systems alone.[57] However, the committee admitted that it lacked the necessary information on which to formulate an educated opinion. USA Digital Radio submitted "at least partial results" for only 18 of the 67 FM lab tests specified in the NRSC's evaluative criteria. "For FM field tests, of the 12 . . . specified in the guidelines, partial results for 5 were submitted. For AM lab tests, of the 25 specified tests, partial results on 8 were submitted. Finally, for the AM field tests, of the 8 specified . . . partial results for 1 were submitted."[58] Lucent Digital Radio's submissions to the NRSC were even paltrier: "For FM lab tests, of the 67 specified in the guidelines, at least partial results were submitted for 5. For FM field tests, of the 12 tests specified . . . partial results for 4 were submitted. For AM lab tests, of the 25 specified . . . partial results on 5 were submitted. Finally, for the AM field tests, of the 8 specified . . . partial results for 0 were submitted."[59] This made it "impossible" for the NRSC to "state unequivocally that . . . IBOC technology provides a significant advance over current analog system performance,"[60] much less provide more granular observations regarding interference tolerance, signal robustness, audio fidelity, or the potential extensibility of new datacasting applications.[61]

CONTENTION BELIES CONSENSUS

On July 12, 2000, USA Digital Radio and Lucent Digital Radio announced their formal merger into a new entity named iBiquity Digital Corporation. Robert Struble, CEO of USADR, became iBiquity's new chief, while LDR president Suren Pai was appointed co-chairman of the board. iBiquity announced plans to combine the two IBOC systems; doing so was expected to shave "a year or two" off the time necessary for the FCC to sanction the technology.[62] Struble exuded confidence: "IBOC is going to be on the air in a year. It's on the air now, but I think it's going to be on the air commercially in a year in dozens, if not hundreds of stations."[63] Between July and December, iBiquity representatives would meet once a month with regulators to push for their blessing of IBOC.[64] iBiquity claimed that FCC endorsement would "remov[e] any uncertainty and accelerat[e] the time when AM and FM broadcasters can join the digital revolution and provide the benefits of digital technology to the American public." There was "no public policy benefit to continuing consideration of new spectrum for DAB when the means exist to reuse the existing spectrum and spectrum shortages threaten the future viability of other wireless services."[65]

More importantly, iBiquity now had a lock on all intellectual property associated with the IBOC protocol. To electronics manufacturers, Struble was blunt: "You guys want to sell radios in 5 or 10 years, you *have* to build IBOCs or you need to work with us" [emphasis in original].[66] Mike Burns, chief engineer at public radio station WAMU in Washington, D.C., remarked that the merger was "a loss for all of us. I just would have loved to see a double-blind competition where the benefits and detriments of each system were exposed . . . That would have been kind of fun."[67] The mood at the 2000 NAB Radio Show was one of uncertainty: *Radio World* editor Paul McLane reported that the gamut of opinion among broadcast managers and engineers ranged from an undeveloped sense of urgency about approving a digital radio system, "magnified by [competition from] satellite radio," to a failure to see any need for digital broadcasting.[68]

In a letter to the editor of *Radio World*, Christopher Maxwell, secretary of the Virginia Center for the Public Press, reiterated the popular sentiment that it was a need for better content—not the adoption of new broadcast technology—which stood to reinvigorate radio. "People don't listen to the Internet because it sounds great; people suffer the Internet audio hassles because the Internet has the programming variety they can't get on the usual FM dial," Maxwell wrote.

And that is why IBOC digital audio broadcasting . . . is going to be a catastrophic failure. IBOC will destroy the listening range of many of the smaller independent commercial, religious and community/college radio stations that serve the "unheard third" of American listeners who are escaping the . . . dial as we speak. To bring them back . . . we need more variety of programming and you get that with more channels.[69]

Scott Todd, a broadcast engineer in Cambridge, Minnesota, told *Radio World* that he "used to be bullish on IBOC, but now I believe it is just bull."[70] Dana Puopolo, the chief engineer of KKBT-FM in Los Angeles, cast the creation of iBiquity as a mechanism by which to maintain the industry's non-competitive status quo.[71]

Contention over IBOC was inflamed in early 2001 when iBiquity announced its technology-licensing scheme.[72] The company initially refused to discuss how much radio transmitter and receiver manufacturers would pay to license IBOC technology, but conceded there would be "an ongoing cost for each unit manufactured." By the end of the year, iBiquity had yet to sign a licensing commitment with any receiver manufacturer,[73] though it did clinch deals with two transmitter companies and three electronics component firms, one of which was an iBiquity investor.[74] Radio stations would also be required to pay licensing fees to broadcast digitally, but initial details were vague. iBiquity proposed various incentives to encourage early adoption of its technology, which typically involved the waiving or capping of certain fees; these included special terms for broadcast conglomerates who were equity investors in the company.[75] By September, iBiquity claimed to have spent "nearly $1 million on developing its technology so far"—a pittance compared to the $100 million XM Satellite Radio budgeted for 2002 on marketing alone.[76]

The year 2001 passed with little policy action on digital radio, but iBiquity and the National Association of Broadcasters kept up the personal pressure on FCC commissioners and senior staff, pushing IBOC as the only viable solution for digital radio in the United States and demanding "prompt FCC action to endorse" the technology.[77] In a PowerPoint presentation to FCC officials in October, iBiquity told the FCC its investors included "14 of the nation's largest broadcasters (including the top 8); 2,300 radio stations with access to 208 million listeners; in Top 50 markets alone, owner stations have a 60% share; owners account for more than half of radio industry revenues." Under a slide entitled, "What Can the FCC Do?" iBiquity was terse and adamant: "Endorse IBOC as the approach US will pursue for digital radio. Discontinue consideration of new spectrum alternatives. Allow stations to begin conversion. Ask for details needed to set IBOC standard."[78] According to FCC Mass Media Bureau Associate Chief Keith Larson, who was repeatedly briefed by iBiquity in private, "the . . . product has tremendous potential," though the agency remained concerned about its potentially deleterious impact on analog radio service.[79]

On November 29, 2001, the National Radio Systems Committee formally endorsed iBiquity's system as the digital framework for U.S. FM radio broadcasting. Like the NRSC's first evaluation, it surfaced as an *ex parte* filing in the FCC's digital radio docket.[80] Wholly based on iBiquity-compiled data, the 215-page report largely parrots the company's own puffery.[81] However, claims of improved audio fidelity were strongly qualified,[82] the potential for analog-to-digital interference was confirmed, and there still wasn't adequate information to grade the feasibility of ancillary services such as datacasting.[83] The FCC subsequently opened a three-month window for public comment on the report.[84]

Industry proponents cast the iBiquity system as the only realistic path toward digital radio in the United States. Their filings included another barrage of NAB-organized astroturf: several parties, ranging from conglomerates to religious networks to stand-alone commercial stations, as well as a leavening of broadcast equipment manufacturers, all cajoled the FCC to "take several steps":

> The Commission should clarify that it is no longer pursuing an out-of-band solution for terrestrial digital radio. iBiquity's demonstration of the viability and benefits of IBOC eliminate the need for examination of alternative approaches. The Commission should also endorse both IBOC as the specific solution for terrestrial digital radio and the iBiquity system. Finally, the Commission should take steps toward the adoption of a formal IBOC standard to encourage broadcasters, receiver manufacturers and consumers to upgrade to digital. This should include proposing rules that enable the introduction of IBOC at the earliest possible date.[85]

The astroturf campaign was inadvertently unmasked by KONP Radio in Port Angeles, Washington, who failed to fill the first blank in the script: "On behalf of [BROADCAST COMPANY X], I am submitting these comments. . . ."[86]

iBiquity and the NAB led this choir. "It has been more than two years since the Commission issued the NPRM," wrote iBiquity. "During that time, there has been tremendous technical and business progress in the development of IBOC . . . Continued Commission silence on the path it will pursue for implementation of DAB . . . will chill the commercial introduction of this innovative and unique technology."[87] Any technical disputes, it claimed, were settled by the NRSC's endorsement; thus, "There should be no further questions about the viability of IBOC technology, its usefulness or the ability to implement IBOC in the field."[88] The NAB was even more insistent, casting the technology as "the radio industry's preferred route to a digital future,"[89] and summed up the status of FM-IBOC in simple terms: "it works; it's ready."[90] Furthermore, the NAB suggested the FCC fold its approval of AM-IBOC into "whatever process is occurring leading to Commission authorization. The time is now ripe and the Commission should act without delay," despite the lack of any substantive testing of the AM system.[91] During the open public comment period, representatives of the trade association met with senior FCC staff to punctuate the need to "expedite" the IBOC rulemaking.[92]

Several iBiquity broadcast investors carried the tune. Susquehanna Radio Corp. commented that while the NRSC report did point to some "shortcomings" in FM-IBOC technology, the system still offered "a definite improvement over existing FM in the United States."[93] Infinity Broadcasting strongly urged the FCC "to endorse and help facilitate rapid implementation of this superior, cutting-edge technology."[94] More to the point, Emmis Communications justified FCC approval of IBOC if only to "giv[e] broadcasters reassurances that their investment in this national technology will be rewarded."[95]

Public broadcasters did not stick to this script. NPR suggested that IBOC would thrive only if it had the potential to provide new program services, not just better audio fidelity.[96] Since iBiquity had not yet developed a multicast capability for FM-IBOC, NPR asserted that the FCC had a duty to proactively mandate such a feature.[97] The Station Resource Group filed comments critical of iBiquity's licensing model, noting, *"Fees that are paid by public broadcasters to iBiquity owners are simply resources that will be displaced from public service programming"* [emphasis in original].[98]

In a surprising change of perspective, the Consumer Electronics Association concurred with the conclusions of the NRSC and urged the FCC to "act swiftly to adopt a single standard for FM IBOC technology."[99] This seemed to signal the trade association's capitulation to the fact that an alternate-band digital radio system was no longer a viable regulatory outcome. However, CEA suggested, "[all] intellectual property included in the standard must either be available free of charge . . . or it must be licensed under reasonable terms in a non-discriminatory manner to anyone who wishes to use it."[100] CEA also reminded the commission to "take note of the fact that many of the commenters in this phase of the proceeding are investors in iBiquity or have entered into special contractual or licensing agreements with iBiquity. While their input to the proceeding is valuable . . . the Commission should also consider the interests of parties who . . . do not currently have an established relationship with iBiquity."[101]

By this time, the majority of independent broadcasters filing comments with the FCC were openly hostile toward IBOC. Anthony Hunt, the general manger of Ball State University's public radio station, expressed concern that iBiquity's system did not represent a meaningful improvement over the analog broadcasting. "In addition to running a station," he wrote, "I also teach college courses. A survey of my class indicated that a majority of my 19-year-old students do not listen to radio. They do not like the music presented to them, so they are looking elsewhere. Now they can find what they want by downloading the music they like, eliminating the need for radio altogether."[102] Hunt believed that IBOC would not address this trend, and unless the FCC found a technology that did "the decision to adopt IBOC will ultimately drive the radio industry into telecommunications obscurity."[103]

The National Federation of Community Broadcasters (NFCB) observed, "IBOC is not the best, and may be the worst [digital radio] system that could be devised." Although it minimized transitional risk for existing radio broadcasters, the technology made "no provision for new stations or new program services. That may be understandable from the viewpoint of the large entities that recently invested many billions to accumulate dozens, or even hundreds of stations in common control. But the FCC is under no duty to follow the investors' risk aversion calculus to its every logical conclusion, where it is regulation itself that has deterred entry and frustrated fresh competition."[104] Aspiring LPFM broadcasters worried about IBOC's effects on their micro-power signals. Duane Wittingham of Macomb, Illinois, one of the owner/managers of WTND-LP, expressed deep concern that "evidence and comments by some organizations that are NOT owners or members of iBiquity [and] the National Association of Broadcasters . . . may turn out to be valid issues and thus could destroy radio reception for many Americans."[105]

The proprietary nature of iBiquity's system also inflamed opposition to it. Radio Kings Bay, Inc., warned of a "potential Commission sanctioned shakedown of our business where no industry pattern for such egregious extortion previously existed,"[106] likened the scheme to "a per user Microsoft software license" (a comparison iBiquity CEO Robert Struble had already made), and blasted iBiquity's business model as "pure sophistry and nothing short of an Enron-caliber attempt to maximize corporate profits without any regard for broadcast operators like RKBI, a last-of-a-dying-breed 'mom and pop' community radio station serving a small South Georgia market."[107]

Several members of the public reiterated the potential for interference that FM-IBOC induced by design.[108] Edward Czelada of Imlay City, Michigan, questioned its long-term viability: "Why should we switch . . . if the audio quality is only subjectively better than analog FM? In 10 years we may have a better idea what is the best method for transmitting digital audio."[109] Others highlighted the apparent imbalance of benefits that would accrue to large broadcast conglomerates under the iBiquity system.[110] Many begged the FCC, at the very least, to engage in independent testing of IBOC instead of relying solely on industry-provided data funneled through the NRSC.[111] Dana J. Woods of Richmond, Virginia, reminded the FCC that "for every letter such as this that you receive, there are probably literally tens of thousands of Americans who are not even aware of this unsavory conversion plan and its implications."[112]

Deep disunity within the broadcast community over the potential benefits of FM-IBOC flared in the trade press. Charles Morgan, senior vice president of Susquehanna Radio and chairman of the NRSC, admitted in

Radio World, "A search of [our evaluation] will show flaws that we wish were not there, but it will also reveal many reasons why broadcasters should embrace this new form of FM broadcasting." However, "For IBOC to succeed, we will need a cooperative effort between iBiquity and broadcasters to get stations on the air quickly . . . if broadcasters do not move forward and support IBOC, it and iBiquity will fail—and if IBOC fails, we, the broadcasters, will also suffer."[113]

iBiquity deflected criticism of its broadcaster licensing fee, portraying it as "consistent with those of other high-end, low-volume software applications and reflects a much smaller component of the licensing revenues iBiquity will receive compared to those from [equipment] manufacturers."[114] The NAB and NRSC took no official position on iBiquity's business model, and *Radio World* reported, "a source close to the Commission said he doubted the agency would 'put a wet blanket on this transition by getting into'" the licensing fee controversy.[115] iBiquity CEO Robert Struble now claimed the company had spent between $100–$150 million in research and development costs, and "we've got to get something back" for that investment.[116]

Concerned that digital radio still lacked a "killer application," National Public Radio formed a Digital Transition Advisory Committee and allocated part of its 2002 digital conversion subsidy from the Corporation for Public Broadcasting to design a multicast feature for iBiquity's FM-IBOC system. NPR also entered into negotiations with iBiquity over an equitable licensing fee structure for stations that might choose to multicast.[117] Mike Starling, the man in charge of NPR's project, was quite candid about the need for it: "The lack of a federal effort for multimode [multicast] digital radio standards could make IBOC a short-lived dead-end and delay the advent of successful digital radio."[118] Thus public broadcasters would assume the role of primary innovators within the iBiquity framework, providing yet another form of subsidy for the privately held technology.

Dissent within the ranks of broadcasters forced some introspection. Microsoft's Skip Pizzi, now part of *Radio World*'s regular roster of commentators, worried, "While controlling one's own destiny is a laudable goal and cherished promise of American enterprise, the push for IBOC has run roughshod over another important American principle: consensus. It thereby violates one of the basic rules of the road in today's technology development. It hearkens back to an earlier, less-enlightened time, when corporate power and oligarchic hegemony could unilaterally control the path and market development of an industry." From its inception, Pizzi claimed IBOC was "fundamentally a blocking policy, primarily intended to retain the status quo for incumbent broadcasters. From an engineering standpoint, it's been a transition plan in search of a technology, with its primary requirements oriented toward damage control rather than growth." He lamented IBOC as an indicator of "how important the business aspects of radio broadcasting are in the United States, and how reduced the public service value of the medium has become."[119]

Although *Radio World* formally came out as an editorial proponent of IBOC in March 2002, editor Paul McLane observed that iBiquity "can appear prickly about questions that I consider legitimate points of discussion" and urged the company to improve its relationship to the industry at large.[120] He also suggested that IBOC's proponents hadn't yet justified wholesale adoption of the technology.[121] These sentiments were a far cry from the unified front of support IBOC's proponents claimed in policy discourse.

FANNING THE FLAMES

Less than a month after public comment had closed on the question of the FCC's potential endorsement of FM-IBOC, iBiquity tendered its own report to regulators on the feasibility of its AM system.[122] The company claimed that its system "offers significant benefits that cannot be matched by analog AM operations . . . The test program also determined that AM IBOC can be introduced without a harmful impact on existing analog AM operations."[123] The National Radio Systems Committee seemed to agree, telling the FCC, "IBOC will transform AM broadcasting through dramatic improvements in AM audio quality . . . Any concerns about [its] potential impact . . . are outweighed by the tremendous benefits IBOC will offer AM broadcasters."[124] Yet its formal evaluation of the technology was mixed: the NRSC observed that implementing AM-IBOC in an interference-free manner was next to impossible, and such interference would most likely be noticed by some listeners.[125] Most notable was the omission of any test data from iBiquity on the performance of an AM-IBOC signal at night, when skywave propagation increased a station's range—and the potential for a digital signal to create destructive interference over a larger area. The NRSC thus recommended that AM-IBOC be implemented during the daytime only.[126]

The FCC opened a third round of public comment to discuss these new findings.[127] iBiquity CEO Robert Struble personally lobbied FCC representatives to produce a Report and Order endorsing the rollout of IBOC on both bands by August 31, 2002.[128] Publicly, iBiquity's bottom-line concern was its bottom line. "Commission endorsement of AM and authorization of digital service, even on an interim basis pending development of final IBOC rules, will foster iBiquity's commercialization schedule and the prompt introduction of the benefits of [this] technology."[129] The National Association of Broadcasters amplified iBiquity's enthusiasm. Although FM-IBOC stood to provide "a qualitative improvement to existing FM quality with few identifiable drawbacks," AM-IBOC, "by contrast, will allow a dramatic—perhaps a transformative—change in AM quality." Such a transformation "may require acceptance of more interference than would be acceptable in a more desirable listening environment, such as FM," but the "constraints on current AM quality" would make this a "tradeoff of limited new interference for vastly improved digital quality."[130]

iBiquity's broadcast-investors were generally enthusiastic in their support of AM-IBOC. Susquehanna Radio recognized that the system was "not without its shortcomings" and explicitly acknowledged the real potential for interference, but concluded, "the gains of the system far outweigh the losses."[131] Cox Radio, Inc., went so far as to ask the commission to implement a plan to phase out analog AM broadcasting in order to alleviate any potential interference concerns.[132] Only Disney, of all major broadcasters expressed a semblance of caution: "To adopt a digital system that creates interference at night in legacy receivers may be the equivalent of burning the bridge you are standing on." That said, "[Disney] also recognizes that, like the sign in the highway construction zone which says 'Temporary Inconvenience, Permanent Improvement,' some compromises are unavoidable during the conversion process."[133] Noting that "virtually every other means of electronic mass media is transitioning to or otherwise deploying digital technology," National Public Radio also endorsed the daytime deployment of AM-IBOC,[134] though it vehemently opposed phasing out analog broadcasting to do it.[135]

The Consumer Electronics Association expressed concerns about AM-IBOC interference,[136] and suggested the FCC mandate that any digitally equipped station be required to mitigate problems it caused to neighboring broadcasters.[137] One consulting engineering firm, Glen Clark & Associates, agreed with the CEA's assessment and openly wondered whether the FCC's AM channel allocation rules provided "sufficient separation between stations" to make IBOC even viable.[138] Many AM station owners argued that implementation should be postponed until the interference potential of the technology was better understood.[139] Neal Newman, the chief engineer of WTTM-AM in Princeton, New Jersey, related his experiences of working at a station that had been an IBOC test bed and questioned the validity of a testing regime partially developed and wholly paid for by iBiquity.[140] Many independent broadcasters simply could not see the economic benefit of deploying AM-IBOC on small-market stations and, like public broadcasters, worried that iBiquity's licensing scheme would make their marginal fiscal situation a dire one.[141]

Radio listeners expressed similarly critical sentiments.[142] Joseph Fela of Plainfield, New Jersey, filed a report cataloguing his reception problems during AM-IBOC field tests in New York.[143] Keven M. Tekel, a citizen with radio engineering experience, tendered his own receiver analysis using eight separate models that illustrated the potential for digital interference on the AM dial.[144] Many members of the public continued to criticize the notion of intentionally degrading incumbent radio service for the benefit of an unproven technology.[145] Others worried about the proprietary nature of iBiquity's business model.[146] A part-time digital radio service, with questionable reception characteristics and audio fidelity, was essentially a nonstarter.[147]

As in previous rounds of policy discourse, IBOC proponents ignored, dismissed, or misconstrued the growing opposition from several quarters.

iBiquity had the temerity to claim, "The record . . . shows that in the past eighteen months IBOC has been transformed from a developing technology into a commercial product awaiting imminent introduction. The public interest will be best served by prompt FCC endorsement of IBOC to support the upcoming launch of commercial receiver sales next year."[148] Commenters who opposed AM-IBOC were characterized as arguing that "AM broadcasters should be relegated to the analog world with no digital future." iBiquity dismissed concerns about the proprietary nature of its technology, confidently noting there was no real commission precedent for the active regulation of its business model.[149] Jumping to the company's defense, Greater Media claimed that the NRSC IBOC test program was "the most thorough program of its kind in the history of radio broadcasting" and guaranteed "a fair, accurate and unbiased evaluation of the system."[150]

Such hubris did not sit well with broadcast engineer Scott Todd: "I'd like to rebut iBiquity's comments that there is overwhelming industry support for IBOC," he told the FCC. "The ones who are so enthusiastic are the corporate suits who think a little technical tinkering is going to reverse declining listenership caused by inept programming . . . I don't see the enthusiasm amongst the rank and file engineers . . . I know that I'll be fighting against it at my company."[151]

iBiquity nonetheless proceeded with its timetable for a public launch. In the first half of 2002, it raised $45 million, mostly from venture capital funds, and projected it would be profitable within a year.[152] The company signed three more licensing deals with transmitter manufacturers,[153] announced its intent to conduct further field-testing on AM-IBOC's nighttime propagation characteristics,[154] and re-branded the technology with the friendly moniker of "HD Radio." According to *Radio World*, "iBiquity's first choice was iDAB," but feared the potential for trademark litigation from Apple Computer.[155]

This upbeat news was offset by growing discontent among broadcasters over HD Radio's fundamental deficiencies. *Radio World* reports from the 2002 NAB Radio Show suggested the number of HD contrarians within the radio industry was on the rise.[156] Further complicating matters, Clear Channel—the largest broadcast conglomerate in the United States and an iBiquity investor—openly questioned the feasibility of HD Radio. Bill Suffa, the company's senior vice president of capital management, told the publication he was not completely sold on the technology. "The whole . . . thing is one of economics. I don't know what they are . . . from a financial basis, it's very difficult to justify going to [HD] at this time." He also expressed concern about iBiquity's licensing fee structure.[157] Reader feedback increasingly questioned in pointed terms whether HD Radio actually represented an improvement over analog, or was instead designed to degrade analog radio so as to make HD *seem* like an improvement.[158] *Radio World* itself chastised iBiquity for not doing more pre-launch public education to increase the potential for marketplace adoption.[159]

Disdain over HD Radio's business plan resonated throughout the industry. The chief engineer of a radio station in Nebraska reported that when iBiquity's licensing scheme was explained at his local Society of Broadcast Engineers meeting, the news broke "like wind when someone cuts the cheese."[160] Skip Pizzi contended that the "$100+ million spent to date on iBiquity's work was expended primarily as a prophylactic investment by commercial broadcasters to ensure that their passage to the digital world would be made on their own terms."[161] He feared that the straightjacket of a proprietary system "will reduce [HD] to insignificance in the consumer marketplace, and thereby cause it to fail . . . The current plan is unbalanced in commercial broadcasters' favor, and is therefore a risk."[162]

In response to this backlash, iBiquity announced a plan to waive the upfront licensing fees for stations that committed to HD Radio broadcasting by the end of 2002 and assured readers of *Radio World* that the company "will work with broadcasters to develop a license fee model that is acceptable to both parties."[163] The publication was unimpressed: "Offering a waiver now, when the biggest groups are the ones that can afford to commit, only advances the perception held by some critics that iBiquity tilts toward its investor partners at the expense of smaller non-investors. Further, this waiver asks managers to commit to ordering equipment before the FCC has even indicated whether it approves of the concept, and in what form."[164]

Personal contacts between regulators and iBiquity grew in frequency as 2002 ground on. Between May and August, senior representatives of the company met at least once a month with FCC officials to press for the unrestricted proliferation of HD Radio.[165] NPR also made two trips to the FCC in September and October to educate commissioners and senior staff about its FM-HD multicast development work.[166] These meetings were integral in wooing regulators to make a leap of faith.

HD RADIO UNLEASHED

On October 10, 2002, the FCC formally selected HD Radio as "the technology that will permit AM and FM radio broadcasters to introduce digital operations efficiently and rapidly." AM and FM stations were given the permission to commence HD broadcasting at their discretion.[167] Formal standard-setting procedures for the technology were deferred to a future rulemaking,[168] yet the policy trajectory could not be clearer: "the Commission will no longer entertain in this proceeding any proposal for digital radio broadcasting other than [HD Radio]."[169] Regulators praised HD as a "remarkable technical achievement," but acknowledged that it could not be implemented "without some service ramifications."[170] The FCC took at face value the comments of "the NRSC and the majority of commenters that the potential for new interference from [HD] operations is insignificant when compared with the advantages and opportunities inherent in this digital technology."[171] It also authorized stations to experiment with multicasting. On the AM band, the FCC heeded the NRSC's advice and did not authorize nighttime HD broadcasting on a blanket basis until further testing was completed.[172]

With regard to iBiquity's licensing model, the FCC cast its decision as an opportunity to assess whether iBiquity and other patent holders "are entering into licensing agreements under reasonable terms and conditions that are demonstrably free of unfair discrimination. The Commission will monitor this situation and seek additional comment as warranted." It characterized the licensing costs provided by iBiquity as not unreasonable "when compared with digital conversion costs in other services" and emphasized the voluntary nature of radio's digital transition.[173]

Three commissioners issued public statements following the unanimous decision. Republicans Kathleen Abernathy and Kevin Martin acknowledged that there might be "some interference with existing services, but we believe that the impact will be minimal and is outweighed by the benefits of digital audio broadcasting."[174] However, it was Democratic commissioner Michael Copps who unwittingly summed up the FCC's predisposition toward "faith-based regulation" on the issue of digital radio. "A few questions remain to be settled, including *how the [HD] system will function in the real world*; what is the potential for and extent of interference that [HD Radio] could cause to existing services; and the technical feasibility of nighttime [AM-HD] transmissions," he remarked [emphasis added]. But Copps expressed confidence that keeping an eye on the pulse of the radio industry would suffice to address any problems that might arise.[175] *Radio World* called the FCC ruling "historic" and reported that the NRSC and iBiquity were in deliberation to fast track a formal HD Radio standard for FCC approval by mid-2003.[176]

The initial policy development of digital radio in the United States was marked by actors primarily speaking past one another, selectively engaging in dialogue relative to their political and economic sway over the proceedings. Legitimate concerns expressed by the public and within the industry were ignored with few exceptions, although the scope and vehemence of public participation was unprecedented for such a technical rulemaking. In simple terms, the aspirations of the public interest in digital radio failed to match up with the limitations of the technology the U.S. radio industry elected to adopt; economic priorities trumped all others, both in the process of constitutive policymaking and in the rhetoric of justification. In the end, iBiquity and its broadcast investors got what they wanted: a technological monopoly on U.S. radio's digital future and permission to begin the monetization process. However, the practical viability of the HD protocol was far from guaranteed.

NOTES

1. Petition for Rulemaking filed by USA Digital Radio Partners, L.P., RM-9395, October 7, 1998.
2. Ibid., 18.
3. Ibid., 7.
4. Notice of USA Digital Radio, RM-9395, January 8, 1999.
5. Federal Communications Commission, *Public Notice*, RM-9395, November 6, 1998.
6. See Comments of the Walt Disney Company, et al., RM-9395, December 23, 1998; Comments of the Gannett Co., RM-9395, December 23, 1998; and Comments of CBS Corporation, RM-9395, December 23, 1998.
7. Comments of Cumulus Media, Inc., RM-9395, December 23, 1998, 5.
8. See Comments of Greater Media, Inc., RM-9395, December 23, 1998; and Comments of the Radio Operators Caucus, RM-9395, December 23, 1998.
9. Comments of National Public Radio, Inc., RM-9395, December 23, 1998.
10. Comments of Clear Channel Communications, Inc., RM-9395, December 23, 1998.
11. Comments of Big City Radio, Inc., RM-9395, December 23, 1998.
12. Reply Comments of Big City Radio, Inc., RM-9395, January 25, 1999.
13. Comments of Ford Motor Company, RM-9395, December 23, 1998.
14. Reply Comments of the Consumer Electronics Manufacturers Association, RM-9395, January 25, 1999.
15. Reply Comments of Citizens' Media Corps, RM-9395, December 22, 1998.
16. See Federal Communications Commission, Media Bureau, Audio Services Division, *In the Matter of Creation of a Low Power Radio Service*, MM 99–25, January 29, 1999.
17. Comments of the Prometheus Radio Project, RM-9395, December 24, 1998.
18. Reply Comments of USA Digital Radio, Inc., RM-9395, January 25, 1999.
19. See *ex parte* Notice of USA Digital Radio, Inc., RM-9395, March 23, 1999; *ex parte* Notice of USA Digital Radio, Inc., RM-9395, April 9, 1999; *ex parte* Notice of USA Digital Radio, Inc., RM-9395, April 9, 1999; *ex parte* Notice of USA Digital Radio, Inc., RM-9395, April 29, 1999; *ex parte* Notice of USA Digital Radio, Inc., RM-9395, May 4, 1999; *ex parte* Notice of USA Digital Radio, Inc., RM-9395, May 4, 1999; *ex parte* Notice of USA Digital Radio, Inc., RM-9395, May 4, 1999; *ex parte* Notice of USA Digital Radio, Inc., RM-9395, May 5, 1999; *ex parte* Notice of USA Digital Radio, Inc., RM-9395, September 3, 1999; *ex parte* Notice of USA Digital Radio, Inc., RM-9395, September 9, 1999; *ex parte* Notice of USA Digital Radio, RM-9395, September 14, 1999; *ex parte* Letter from USA Digital Radio, RM-9395, September 22, 1999; *ex parte* Notice of USA Digital Radio, Inc., RM-9395, September 23, 1999; *ex parte* Notice of USA Digital Radio, Inc., RM-9395, September 29, 1999; *ex parte* Notice of USA Digital Radio, Inc., RM-9395, September 29, 1999; *ex parte* Notice of USA Digital Radio, Inc., RM-9395, September 28, 1999; *ex parte* Notice of USA Digital Radio, Inc., RM-9395, October 1, 1999; and *ex parte* Notice of USA Digital Radio, Inc., RM-9395, October 7, 1999.
20. Steve Behrens, "Field Testing Resumes for Radio's Digital Best Hope," *Current* XVIII, no. 15 (August 16, 1999), 1, 19.
21. See Federal Communications Commission, Mass Media Bureau, Audio Services Division, *Notice of Proposed Rulemaking: Digital Audio Broadcasting Systems and Their Impact on the Terrestrial Radio Broadcast Service*, (MM) Docket 99–325, November 1, 1999.

22. Ibid., 2.
23. Ibid., 8–9.
24. Ibid., 21–22.
25. Submission of Report of USA Digital Radio, Inc., MM 99–325, December 21, 1999.
26. Ibid., 4.
27. Comments of Lucent Digital Radio, Inc., MM 99–325, January 24, 2000.
28. Ibid., 17–19.
29. See Comments of WTRW Incorporated, MM 99–325, January 10, 2000; Comments of KMSO-FM, Rochester Radio Group, and KSOB/KSQB-FM, MM 99–325, January 11, 2000; Comments of Resort Radio Systems, MM 99–325, January 11, 2000; Comments of KOFO-AM, MM 99–325, January 13, 2000; Comments of WAJI-FM and WLDE-FM, MM 99–325, January 13, 2000; Comments of KOOL 103.5 WASE, MM 99–325, January 13, 2000; Comments of Cox Radio, Inc., MM 99–325, January 20, 2000; Comments of Perception Media Group, Inc., MM 99–325, January 24, 2000; Comments of Gulf Coast Broadcasting Company, Inc., MM 99–325, January 24, 2000; and Comments of Wright Broadcasting Systems, Inc., MM 99–325, January 24, 2000.
30. See Comments of Chase Capital Partners, MM 99–325, January 24, 2000; Comments of Susquehanna Radio Corp., MM 99–325, January 24, 2000; Comments of Hispanic Broadcasting Corporation, MM 99–325, January 24, 2000; Comments of Cox Radio, Inc., MM 99–325, January 24, 2000; Comments of Gannett Co., Inc., MM 99–325, January 24, 2000; Comments of Entercom Communications Corporation, MM 99–325, January 24, 2000; Comments of Infinity Broadcasting Corporation, MM 99–325, January 24, 2000; Reply Comments of Radio One, Inc., MM 99–325, February 22, 2000; and Comments of the National Association of Broadcasters, MM 99–325, January 24, 2000.
31. Comments of USA Digital Radio, Inc., MM 99–325, January 24, 2000.
32. Submission of Report of USA Digital Radio, 3.
33. Ibid., 4.
34. Ibid., 1–2, 5.
35. NRSC DAB Subcommittee, "Goals and Objectives," as adopted May 14, 1998, in Comments of the National Association of Broadcasters, MM 99–325, January 24, 2000, 1.
36. NRSC DAB Subcommittee, "In-band/On-channel (IBOC) Digital Audio Broadcasting (DAB) System Evaluation Guidelines" adopted April 17, 1999, revised May 25, 1999, 12–13.
37. Comments of National Public Radio, Inc., MM 99–325, January 24, 2000, 2, 6.
38. See Ibid., 2–3, 6, and Reply Comments of National Public Radio, Inc., MM 99–325, February 22, 2000.
39. Comments of Sony Electronics Inc., MM 99–325, January 24, 2000, 2.
40. Ibid., 2–3.
41. Comments of Visteon Automotive Systems, MM 99–325, January 24, 2000, i, 1, 4, 10–11.
42. Ibid., 11.
43. Comments of the Consumer Electronics Association, MM 99–325, January 24, 2000, ii-iii.
44. Ibid., ii, 4–10, 13–14.
45. See Comments of the Consumer Electronics Association, Appendix B, "Concept Design for a Mobile Multimedia Broadcast Service," MM 99–325, January 24, 2000.

46. Reply Comments of David P. Maxson, MM 99–325, February 23, 2000, 19–20.
47. Ibid., 24.
48. Comments of Clarkston Broadcasters, Inc., MM 99–325, December 29, 1999.
49. See Comments of David S. Forsman, MM 99–325, December 15, 1999, 2; Comments of Ted M. Coompan, et. al., MM 99–325, January 24, 2000, 4; Comments of Pete Tridish, MM 99–325, January 24, 2000, 2, 4–5; and Reply Comments of REC Networks, MM 99–325, February 14, 2000, 4.
50. Reply Comments of the Virginia Center for the Public Press, MM 99–325, March 21, 2000, 8, 27.
51. See Comments of Blair Alper, MM 99–325, December 23, 1999; and Comments of Virginia Center for the Public Press, MM 99–325, January 27, 2000, 6.
52. See Comments of David S. Forsman, 2; Comments of REC Networks, MM 99–325, January 6, 2000; Comments of Virginia Center for the Public Press, 13–17; Comments of Robert Bornkamp, MM 99–325, January 27, 2000; Comments of Pete Tridish, 6–7; and Reply Comments of REC Networks, 1.
53. Comments of Ted M. Coompan, et al., 2–3.
54. Reply Comments of USA Digital Radio, Inc., MM 99–325, February 22, 2000, 8.
55. Reply Comments of Lucent Digital Radio, Inc., MM 99–325, February 22, 2000, 7–8.
56. See ibid., 19, and Reply Comments of the National Association of Broadcasters, MM 99–325, January 22, 2000, 5–6.
57. NRSC DAB Subcommittee, "Evaluation of USA Digital Radio's Submission to the NRSC DAB Subcommittee of Selected Laboratory and Field Test Results for its FM and AM Band IBOC System," 6, in *ex parte* Notice by the National Association of Broadcasters, MM 99–325, May 12, 2000.
58. Ibid., 7.
59. NRSC DAB Subcommittee, "Evaluation of Lucent Digital Radio's Submission to the NRSC DAB Subcommittee of Selected Laboratory and Field Test Results for its FM and AM Band IBOC System," 7, in *ex parte* Notice by the National Association of Broadcasters, MM 99–325, May 12, 2000.
60. Ibid., 8.
61. Ibid., 10, 13–22.
62. Mike Janssen, "Technical Merger May Speed Adoption of Digital Radio," *Current* XIX, no. 13 (July 17, 2000), 5.
63. "NAB Endorses IBOC; Feels Infinity Loss," *Radio World*, July 18, 2001, 2.
64. See *ex parte* Letter from USA Digital Radio, Inc., MM 99–325, July 28, 2000; *ex parte* Letter from Lucent Digital Radio, MM 99–325, August 2, 2000; *ex parte* Letter from iBiquity Digital Corporation, MM 99–325, September 19, 2000; *ex parte* Letter from iBiquity Digital Corporation, MM 99–325, October 13, 2000; and *ex parte* Letter from iBiquity Corporation, MM 99–325, November 1, 2000.
65. *Ex parte* Letter from iBiquity Digital Corporation, MM 99–325, December 11, 2000, 3–4.
66. Quoted in Leslie Stimson, "Robert Struble Steers New IBOC DAB Race," *Radio World*, September 13, 2000, 6–7.
67. Quoted in Janssen, "Technical Merger May Speed Adoption of Digital Radio."
68. Paul J. McLane, "From the Editor: Postcard From San Francisco," *Radio World*, October 11, 2000, 4.
69. Christopher Maxwell, "Don't Reinvent the Wheel," *Radio World*, November 22, 2000, 86.
70. Scott Todd, "Rebuttal," *Radio World*, December 20, 2000, 62.
71. Dana Puopolo, "Grand Alliance?", *Radio World*, November 22, 2000, 4.
72. "iBiquity Integrates Staff, Board," *Radio World*, January 17, 2001, 5.

73. Leslie Stimson, "iBiquity, Hyundai Team Up," *Radio World*, September 12, 2001, 5.
74. Leslie Stimson, "A DAB Call For Action," *Radio World*, September 26, 2001, 3, 8.
75. Leslie Stimson, "Reaction to IBOC Fees," *Radio World*, May 23, 2001, 3.
76. Leslie Stimson, "A DAB Call For Action," 3.
77. See *ex parte* Letter from iBiquity Digital Corporation, MM 99–325, March 3, 2001; *ex parte* Letter from iBiquity Corporation, MM 99–325, April 18, 2001; *ex parte* Letter from the National Association of Broadcasters, MM 99–325, June 27, 2001; *ex parte* Letter from the National Association of Broadcasters, MM 99–325, July 16, 2001; *ex parte* Letter from iBiquity Digital Corporation, MM 99–325, June 28, 2001; and Leslie Stimson, "FM IBOC Results Scrutinized," *Radio World*, September 12, 2001, 1, 5–6.
78. *Ex parte* Letter from iBiquity Digital Corporation, MM 99–325, October 9, 2001.
79. Leslie Stimson, "IBOC Step in November?" *Radio World*, September 26, 2001, 3.
80. Submission of Report of NAB and CEA (on Behalf of the NRSC), MM 99–325, December 3, 2001.
81. Ibid., 6.
82. Ibid., 8, 27.
83. Ibid., 47.
84. Federal Communications Commission, Mass Media Bureau, *Public Notice*, MM 99–325, December 19, 2001.
85. See Comments of Evangelistic Alaska Missionary Fellowship, MM 99–325, February 12, 2002; Comments of KMAT, MM 99–325, February 19, 2002; Comments of Radio Station WAWL, MM 99–325, April 10, 2002; Comments of LA Media, MM 99–325, March 25, 2002; Comments of Journal Broadcast Corporation, MM 99–325, February 19, 2002; Comments of Clear Channel Communications, Inc., MM 99–325, February 19, 2002; Comments of Pinwheel, Inc., MM 99–325, February 19, 2002; Comments of Maumee Valley Broadcasting Inc., MM 99–325, February 19, 2002; Comments of Shively Labs, MM 99–325, February 7, 2002; Comments of Nautel Maine, Inc., MM 99–325, February 15, 2002; Comments of JVC, Inc., MM 99–325, February 19, 2002; and Comments of Toko America, MM 99–325, February 19, 2002.
86. Comments of KONP, MM 99–325, March 1, 2002.
87. Comments of iBiquity Digital Corporation, MM 99–325, February 20, 2002, 4.
88. Ibid., 15.
89. Comments of the National Association of Broadcasters, MM 99–325, February 19, 2002, i.
90. Ibid., 2.
91. Ibid., ii.
92. *Ex parte* Letter from the National Association of Broadcasters, MM 99–325, February 6, 2002.
93. Comments of Susquehanna Radio Corp., MM 99–325, February 19, 2002, 2.
94. Comments of Infinity Broadcasting Corporation, MM 99–325, February 19, 2002, 2, 5.
95. Comments of Emmis Communications, MM 99–325, February 19, 2002.
96. Reply Comments of National Public Radio, MM 99–325, March 21, 2002, 8–10.
97. Ibid., 5.
98. Comments of Station Resource Group, MM 99–325, March 21, 2002.
99. Comments of the Consumer Electronics Association, 1–2.
100. Comments of the Consumer Electronics Association, 3.

101. Reply Comments of the Consumer Electronics Association, 3–4.
102. Comments of Anthony Hunt, MM 99–325, February 22, 2002.
103. Ibid.
104. Reply Comments of the National Federation of Community Broadcasters, MM 99–325, March 29, 2002.
105. Reply Comments of Duane Wittingham, MM 99–325, March 20, 2002; see also Comments of Chuck Conrad, MM 99–325, February 20, 2002.
106. Reply Comments of Radio Kings Bay, Inc., MM 99–325, March 20, 2002, 1–2.
107. Ibid., 2.
108. See Comments of the Amherst Alliance, MM 99–325, February 19, 2002; Comments of Matthew A. Hayes, MM 99–325, February 28, 2002; and Comments of John M. Roberts, MM 99–325, March 21, 2002.
109. Comments of Edward Czelada, MM 99–325, March 21, 2002, 3.
110. Comments of Donald E. Niccum, MM 99–325, March 14, 2002, 1.
111. See Reply Comments of the Virginia Center for the Public Press, 8–9; Comments of Caroline Cox, MM 99–325, March 21, 2002; and Comments of Barbara Reuther, MM 99–325, March 21, 2002.
112. Comments of Dana J. Woods, MM 99–325, March 20, 2002; see also Reply Comments of Ted M. Coopman, MM 99–325, March 21, 2002, 2.
113. Charles T. Morgan, "IBOC: What Engineers Should Know," *Radio World*, February 1, 2002, 6.
114. Scott Stull, "Licensing Fees To Be Levied," *Radio World*, April 10, 2002, 8.
115. Leslie Stimson, "IBOC Fees Stir Reaction," *Radio World*, April 10, 2002, 12.
116. Leslie Stimson, "The 'Official Launch' of IBOC DAB," *Radio World*, May 8, 2002, 12.
117. Mike Janssen, "Pubradio Drafts Wish List as IBOC Standard Nears," *Current* XXI, no. 6 (March 25, 2002), 1, 14, 16.
118. Mike Starling, "Digital Radio *Could* be More Satisfying, But Will It?," *Current* XXI, no. 6 (March 25, 2002), 17, 21.
119. Skip Pizzi, "FMX+RBDS = IBOC," *Radio World*, March 1, 2002, 16.
120. Paul McLane, "From the Editor: These Pages Are Open to All," *Radio World*, March 27, 2002, 4.
121. Editorial, "The Means, Not the Ends," *Radio World*, March 1, 2002, 54.
122. Report of iBiquity Digital Corporation, MM 99–325, April 15, 2002.
123. Ibid., i.
124. Ibid., 18, 28.
125. Report of the NRSC Evaluation of AM IBOC, as submitted by the National Association of Broadcasters et al., MM 99–325, April 16, 2002, 8, 11, 24, 54, 57.
126. Ibid., 8–9.
127. Federal Communications Commission, Mass Media Bureau, *Public Notice*, MM 99–325, April 19, 2002.
128. *Ex parte* Comments of iBiquity Corporation, MM 99–325, April 4, 2002, 17.
129. Comments of iBiquity Digital Corporation, MM 99–325, June 18, 2002, 12–13.
130. Comments of the National Association of Broadcasters, MM 99–325, June 18, 2002, 10.
131. See Comments of Susquehanna Radio Co., MM 99–325, June 17, 2002, 2–3; and Reply Comments of Susquehanna Radio Co., MM 99–325, July 18, 2002, 3.
132. Comments of Cox Radio, Inc., MM 99–325, June 18, 2002, 2–3.
133. Reply Comments of the Walt Disney Company and ABC, Inc., MM 99–325, July 18, 2002, 6–7.

134. Comments of National Public Radio, MM 99–325, June 18, 2002.
135. Reply Comments of National Public Radio, Inc., MM 99–325, July 18, 2002.
136. Reply Comments of the Consumer Electronics Association, MM 99–325, July 18, 2002.
137. Ibid, 4–5.
138. Comments of Glen Clark & Associates, MM 99–325, July 18, 2002), 2.
139. See Comments of Bob Carter, WGAI-AM, MM 99–325, June 18, 2002; and Reply Comments of James Crystal Enterprises, LLC, MM 99–325, July 18, 2002.
140. Comments of Neal Newman, MM 99–325, June 11, 2002.
141. See Comments of Scott A. Todd, MM 99–325, June 17, 2002; and Comments of WGUF-FM, Inc., MM 99–325, June 13, 2002.
142. See Comments of Richard W. Kenneally, MM 99–325, April 26, 2002; Comments of REC Networks, MM 99–325, April 29, 2002, 3; Comments of Various citizens, MM 99–325, April 26, 2002; Comments of John Pavlica, Jr., MM 99–325, May 21, 2002; Comments of Thomas Bryant, MM 99–325, June 10, 2002; Comments of Kit Sage, MM 99–325, June 10, 2002; Comments of Harry L. Helms, MM 99–325, June 10, 2002, 1; Comments of "Citizen from Novi Michigan," MM 99–325, June 12, 2002; Comments of Bruce A. Conti, MM 99–325, June 13, 2002; Comments of Christopher Cuff, MM 99–325, June 17, 2002; Comments of Glenn Hauser, MM 99–325, June 17, 2002; Comments of Michael Bugaj, MM 99–325, June 17, 2002; Comments of Kevin M. Tekel, MM 99–325, June 17, 2002; Comments of Robert Meuser, MM 99–325, June 17, 2002; Comments of Robert Foxworth, MM 99–325, June 18, 2002; Comments of Ian R. Davidson, MM 99–325, June 18, 2002; Comments of Charles Hutton, MM 99–325, June 18, 2002; Comments of Michael J. Richard, MM 99–325, June 18, 2002; and Reply Comments of Marv C. Southcott, MM 99–325, July 18, 2002.
143. Comments of Joseph Fela, MM 99–325, June 17, 2002.
144. Comments of Kevin M. Tekel, MM 99–325, June 18, 2002.
145. See Comments of Jay Rogers, MM 99–325, June 11, 2002; Comments of Russell Skadl, MM 99–325, June 18, 2002, 2; Comments of Juan C. Gualda, MM 99–325, June 18, 2002; and Reply Comments of John Pavlica, Jr., MM 99–325, July 18, 2002, 1.
146. See Comments of REC Networks, 3; Comments of David S. Forsman, MM 99–325, June 3, 2002, 1; Comments of John J. Rieger, MM 99–325, June 17, 2002, 1; Comments of Russell Skadl, 1; Comments of Lawrence Waldbillig, MM 99–325, June 19, 2002, 1; Reply Comments of John Pavlica, Jr., 2; and Reply Comments of Robert A. Meuser, MM 99–325, July 30, 2002, 1.
147. See Comments of J. S. Gilstrap Jr., MM 99–325, June 18, 2002, 3; and Reply Comments of Kevin Redding, MM 99–325, July 22, 2002, 1.
148. Reply Comments of iBiquity Digital Corporation, MM 99–325, July 18, 2002, 1; see also Reply Comments of the National Association of Broadcasters, MM 99–325, July 18, 2002, 2.
149. Reply Comments of iBiquity Digital Corporation, 6.
150. Reply Comments of Greater Media, Inc., MM 99–325, July 18, 2002, 3.
151. Reply Comments of Scott A. Todd, MM 99–325, July 18, 2002.
152. "iBiquity Raises $45M," *Radio World*, May 22, 2002, 2, 14.
153. See Editorial, "IBOC Exciter-ment," *Radio World*, May 22, 2002, 62; and Dale Mowry, "Time for a Shared IBOC Vision," *Radio World*, July 17, 2002, 5.
154. See "iBiquity Testing on WOR(AM)," *Radio World*, August 14, 2002, 2, 10; and Leslie Stimson, "Ibiquity Plans AM Night Tests," *Radio World*, September 1, 2002, 3.

155. Leslie Stimson, "It's Not IBOC, But 'HD Radio'," *Radio World*, September 1, 2002, 3.

156. See Skip Pizzi, "Concerns About IBOC Grow," *Radio World*, May 22, 2002, 17–18; and Guy Wire, "A Bumpy Ride for Ibiquity in Vegas," *Radio World*, May 22, 2002, 28.

157. Leslie Stimson, "Suffa Eyes IBOC With Caution," *Radio World*, August 1, 2002, 1, 8, 10.

158. See Glenn Finney, "Digital Radio Wars," *Radio World*, June 19, 2002, 53; and Gordon Carter, CPBE, Chief Engineer, "IBOC and Classical Music," *Radio World*, July 17, 2002, 54.

159. Editorial, "IBOC Is Now 'HD Radio,'" *Radio World*, September 1, 2002, 78.

160. Mike Shane, "More on IBOC," *Radio World*, June 19, 2002, 53.

161. Skip Pizzi, "IBOC DAB, in the Public Eye," *Radio World*, June 19, 2002, 14, 22.

162. Skip Pizzi, "Concerns About IBOC Grow," *Radio World*, May 22, 2002, 18.

163. Scott Stull, "Licensing Fee Waiver Demystified," *Radio World*, October 23, 2002, 6.

164. Editorial, "The IBOC Fee Waiver," *Radio World*, October 9, 2002, 54.

165. See *Ex parte* Comments of iBiquity Digital Corporation, MM 99–325, May 20, 2002, 1; *ex parte* Comments of iBiquity Digital Corporation, MM 99–325, June 17, 2002; *ex parte* Comments of iBiquity Digital Corporation, MM 99–325, May 3, 2002, 2; *ex parte* Notice of iBiquity Digital Corporation, MM 99–325, July 23, 2002; and *ex parte* Notice of iBiquity Digital Corporation, MM 99–325, August 12, 2002.

166. See *ex parte* Notice of National Public Radio, Inc., MM 99–325, September 30, 2002; and *ex parte* Notice of National Public Radio, Inc., MM 99–325, October 2, 2002.

167. Federal Communications Commission, Mass Media Bureau, *Report and Order*, MM 99–325, October 10, 2002, 1.

168. Ibid., 13–14.

169. Ibid., 17.

170. Ibid., 6.

171. Ibid., 12.

172. Ibid., 9–12.

173. Ibid., 13.

174. Joint Statement of Commissioners Kathleen Q. Abernathy and Kevin J. Martin, in ibid.

175. Statement of Commissioner Michael J. Copps, in ibid.

176. Stimson, "FCC Okays 'Historic' IBOC Order," *Radio World*, October 23, 2002, 1, 6.

5 The Troubled Proliferation of HD Radio

With the FCC's blessing, HD Radio's proponents may have thought they were free and clear to implement the technology with few restrictions, and that its uptake would be a given. However, as stations began to implement HD Radio and iBiquity Digital Corporation refined its system, its fundamental faults became increasingly clear. This further alienated independent broadcasters and the public, who were already of the mind that such a questionable and proprietary technology threatened the viability of terrestrial radio itself.

STUMBLING OUT OF THE GATE

Shortly after the FCC's 2002 decision, radio listeners filed reports with regulators detailing HD-related interference. Commenters from Arizona to Florida told the FCC about "rushing water" noises caused by digital hash from AM-HD stations in New York, Cincinnati, and Chicago that had ruined nighttime reception of AM signals in their communities.[1] Other members of the public noted that neither iBiquity, the National Radio Systems Committee, nor the FCC had adequately considered the diversity of radio receivers that existed in the marketplace, and warned that many inexpensive models would suffer significant HD-related interference.[2]

This did not stop the technology's proponents from pushing ahead with their rollout. The Corporation for Public Broadcasting hired consulting engineer Doug Vernier to help plan NPR's digital transition; WUSF-FM in Tampa, Florida, became the first NPR affiliate to place an order for an HD transmitter,[3] and the CPB announced it would make $4.5 million available to qualified stations to help cover digital conversion costs.[4] Among commercial broadcasters, 35 radio station group-owners announced plans to implement HD Radio on some 300 stations in 40 markets; most of the companies involved had an equity stake in iBiquity.[5] CEO Robert Struble told *Radio World* in January 2003 that he expected the FCC to formalize HD Radio as the nation's digital radio standard by the end of year,[6] and that the number of HD-compatible stations on the air would double by the end of 2004.[7]

The company also announced another licensing incentive program, slashing one-time broadcaster fees for early adopters; noncommercial stations that agreed to implement HD before mid-year would have the initial license fee waived completely.[8]

These seemingly positive developments were not reflected in the commentary of working broadcasters in the trade press. KRCO-AM operations director Mike Shane called the imposition of HD Radio "a spectrum grab by those currently in possession of frequencies to keep future competition off the air . . . I am ashamed to have missed this obvious point up till now and even more ashamed to be part of this business that, with complicity from the FCC, is perpetrating a major scam on the American listening public."[9] More broadcasters reported their own HD-related interference experiences on both the AM and FM bands.[10] Many complained that the combination of station equipment upgrades necessary to implement HD Radio and iBiquity's licensing plan would make digital conversion unaffordable for most small- to medium-market broadcasters.[11] *Radio World* observed that HD Radio "pits large group owners against smaller station owners and splits NAB's radio membership."[12] Editor Paul McLane was shaken by the rising tide of skepticism.[13] He expressed concern about the potential for signal interference to grow as more digital broadcasters took to the air. "It is hard for us to stand by and ignore the obliteration of any station on the dial due to interference. But we are hoping for the best the technology has to offer and are looking toward the future as a larger picture."[14]

The NRSC threw that future into disarray when it voted to suspend standard-setting work on the HD Radio protocol in May 2003. Following the technology's preliminary endorsement by the FCC, iBiquity "upgraded" its system by replacing the algorithm designed to encode HD audio.[15] Unfortunately, this new codec performed so poorly—especially at low-bit rates like those used with AM-HD and FM-HD multicasting—that the entire viability of the system was called into question. The consensus of NRSC members was that the system's audio quality was now simply "unacceptable."[16]

Radio World minced no words about this pitfall. The NRSC standards-setting process was characterized as "far too secretive, and not well-defined," according to editor Paul McLane. "Why do we find ourselves so far along on the HD Radio rollout with a core piece of the digital radio puzzle not yet firmly in place? . . . [T]he NRSC should open its meetings to all concerned parties—including those of us who cover the industry on your behalf . . . Public debate and news coverage will ensure vigorous scrutiny, to everyone's benefit."[17] A re-think of the system was required, because "the NRSC could not proceed and remain legitimate." At least one transmitter manufacturer with an equity investment in iBiquity halted the construction of HD-compatible units until the problem could be sorted out.[18]

It took three months for iBiquity to replace its failed codec. During this period of retrenchment, the company laid off Project Acorn facilitator and USA Digital Radio cofounder E. Glynn Walden, along with two other managers and 32 line-level employees.[19] Milford Smith, chairman of the NRSC's DAB Subcommittee, said he was "incredibly shocked and disappointed" by the layoffs and believed Walden had represented "the most reliable and dependable contact" at iBiquity. *Radio World* reported that the personnel cuts were due to the codec controversy as well as the company's cash burn-rate, estimated to be $2.1 million per month. Although the publication reported that iBiquity raised "roughly $100 million" in a recent round of financing, direct investment from broadcasters had dropped significantly, and the shortfall had to be made up by overtures to venture capitalists.[20] iBiquity CEO Robert Struble denied the company was in a fiscal squeeze, although *Radio World* reported that it was quietly seeking more broadcast investors beyond the national conglomerates, with little success.[21]

Broadcasters sounded off in the trade press with increasing disdain. Commentator Skip Pizzi deplored the fact that HD Radio remained in "such an immature, unstable and speculative state" and worried that iBiquity's layoffs gave good reason for the industry to question the "corporation's ongoing viability" and "management sensibility."[22] More reports arrived at *Radio World* detailing experiences with HD-induced interference.[23] Some broadcasters openly called on the FCC to re-open its exploration of digital radio to include other systems.[24] NRSC chairman Charles Morgan openly worried about the shaky rollout: "Without broadcasters taking the lead and investing in installations, I fear that [HD] may become stalled or simply die a slow death. If that happens, I believe we will see some form of new-band terrestrial radio . . . Existing broadcasters will have no assurances that they will have the same role in this new service as they have today."[25] This sentiment re-emphasized the notion that HD proponents were not out to develop a technology that worked well, but rather to lock in a process for radio's digital transition that would preserve their place of primacy on the public airwaves.

Yet as HD Radio's proliferation looked increasingly questionable in the real world, its proponents became more emphatic about its necessity in the policy world. They creatively re-contextualized the FCC's desire to implement "further rules" governing the technology into a crusade to push the agency toward "final rules" deciding the question of HD Radio's exclusivity once and for all. They did not blame the technology's fundamental deficiencies, anemic listener interest, or iBiquity's business model for its slow uptake, but "regulatory uncertainty," which only the FCC could fix by laying to rest any doubts about the technology's viability.[26] However, very few supporters favored a mandatory analog/digital conversion deadline. Most corporate proponents along with 47 state broadcasters' associations urged the FCC to allow marketplace forces to govern the transition, on the premise that analog broadcasting would remain "the mainstay of radio . . . for years to come."[27]

Public broadcasters' primary goal in the push for "final rules" was to cement multicasting as a standard feature in the HD system. NPR organized nearly 150 of its affiliate stations, the educational institutions they are wedded to, and loyal listeners to deluge the FCC with positive sentiments about their experience with or belief in the potential of multicasting to increase program diversity.[28] Both NPR and CPB framed multicasting as "the driving force" behind HD Radio's success, and the only mechanism by which listener experience would be significantly improved in a digital broadcast environment.[29] NPR would intensify its public campaign in a series of *ex parte* meetings with senior FCC staff, emphasizing the need for multicasting to be required by rule.[30]

Independent broadcasters reacted with disdain to the push for "final rules." They accused the FCC of being "enamored with all things digital" and failing to adequately address HD's shortfalls.[31] Many FM broadcasters lamented the fact that not enough data had been collected to guarantee even the most basic interference protections to analog service.[32] Some provided further analyses of existing FM-HD interference.[33] With regard to AM-HD, they pulled no punches. Oklahoma-based Reunion Broadcasting noted that although there were just some 30 AM-HD stations broadcasting nationwide by mid-2004, they were causing enough harm on the band to raise serious concerns.[34] The potential for interference from unrestricted AM-HD operations terrified many small broadcasters,[35] who pleaded with the FCC to carefully examine this problem.[36] Many broadcasters noted an explosion of digitally induced "white noise," "hissing," and "buzzing" on the AM dial,[37] and told the FCC that they would not adopt iBiquity's technology due to its monopolistic licensing system.[38]

Members of the public redoubled their efforts to explain to regulators that not only would HD Radio be detrimental to listener choice, but that they simply didn't want it. Several media reform organizations, under the moniker of the "Public Interest Coalition," took umbrage with the idea that HD signals used "no new spectrum." "This technology . . . allows broadcasters to engage in activity which is the equivalent of constructing additional buildings on their spectrum sidewalks, taking space they have not been previously allowed to use." Given "a media environment increasingly characterized by concentrated ownership, commercially-driven content, and a lack of civic engagement," they suggested the FCC use its HD rulemaking to "represent a new stage in the ongoing evolution of the public interest standard: a needed reassessment in light of dramatic changes in communications technology, market structures, and the needs of a democratic society."[39]

Interference concerns and HD Radio's proprietary nature also remained major points of contention with radio listeners. More than a dozen individuals filed comments opposing the full-time expansion of AM-HD broadcasting, including reports detailing the reception problems that the handful of AM-HD stations on air had already caused.[40] The intimate fiscal ties between iBiquity and the nation's largest radio conglomerates, argued the Amherst Alliance, "creates for a number of broadcasters an inherent conflict-of-interest that makes them questionable judges of which Digital Radio technology is really best for radio."[41] The FCC's singular reliance on industry data was clearly designed "to further the interests of the entities that submit it," according to the Public Interest Coalition. "The data is not compiled in a manner that would aid independent academic research. Because most of the data presently relied upon by the FCC is proprietary and thus unavailable for others to use," there was no honest way to quantify HD Radio's promised benefits.[42]

ibiquity called the growing concerns of HD-related interference "overblown" and argued that its "exhaustive testing" of the technology obviated the need to discuss the issue.[43] The NAB singled out complaints of "mercurial" AM-HD interference only to dismiss them as "theoretical or analytical engineering concerns."[44] National Public Radio joined this rebuttal, asserting that opening up digital radio signals to new broadcast-entrants would require "a sweeping new regulatory regime . . . that would, in fact, undermine the DAB transition and [is], therefore, contrary to the public interest."[45]

By mid-2004, only 100 HD-enabled stations were on the air nationwide, prompting iBiquity to hire marketing help to promote the technology's uptake by broadcasters.[46] The company and its investors continued to produce a steady stream of news that gave HD Radio some semblance of momentum. Clear Channel announced it would spend up to $100 million over the next decade to convert its stations to HD, though the decision was spurred by undisclosed "incentives" that significantly reduced the company's license fee burden.[47] iBiquity touted a "historic agreement" with 21 radio group owners to convert 2,000 stations in the top 100 radio markets by the end of 2005; Entercom and Greater Media pledged to convert *all* their stations within a matter of months.[48] "Industry observers" predicted broadcasters would spend "approximately $115 million . . . over the next few years" to upgrade their stations to digital.[49] *Radio World* found a consulting engineer who predicted HD Radio would reach marketplace criticality by 2009.[50]

When a radio station in Puerto Rico went HD in 2005, iBiquity celebrated the occasion as "position[ing] the technology for widespread adoption outside the continental U.S."[51] The company raised another $30 million in financing, mostly from venture capital firms, bringing iBiquity's total equity investment to $135 million since its founding in 2000.[52] The Corporation for Public Broadcasting reported that 50 public radio stations were broadcasting in some form of HD and 262 station-conversion projects were underway, at a cost of some $20 million in taxpayer subsidies.[53] NPR encouraged its FM affiliates to petition the FCC for multicasting authority, so as to demonstrate industry willingness to adopt the feature.[54] NPR also unveiled plans to offer five "turnkey" music formats to fill multicast programming without the need for stations to invest in more local programming resources.[55]

That was the good news. In May 2004, *Radio World* reported that some AM stations had turned off their digital sidebands due to interference concerns; iBiquity claimed these cases were aberrations.[56] The Canadian Association of Broadcasters and Canadian Broadcasting Corporation asked its own regulators to "notify the FCC that the authorization of [AM-HD] transmissions, especially at night, will very likely result in harmful interference to Canadian signals."[57] *Radio World* continued to receive a steady stream of letters from broadcasters documenting AM-HD interference in their communities.[58]

Some early-adopters took to the trades attempting to assuage industry discontent. Thomas R. Ray III, chief engineer of AM-HD station WOR in New York, admitted that radio's digital transition would take upward of 20 years and required compromises to analog signals, but he argued that any other outcome was untenable. "[HD Radio] isn't the three-headed monster it's been portrayed to be . . . Remaining the same while the world marches past us will place terrestrial AM and FM broadcasting among the dinosaurs, rendering us irrelevant."[59] *Radio World*'s anonymous engineering commentator, "Guy Wire," latched onto the "naysayer" moniker to portray HD critics as Luddites:

> Just because it's digital doesn't mean it's better, they say . . . This reminds us all of the horse-and-carriage fans of the 1890s . . . HD Radio right now is like the first automobiles to travel unpaved roads. Think about what it will offer in another five or 10 years. History is squarely on the side of better technology as it pushes aside older, less efficient methods . . . It's really very simple. The digital bus with HD Radio onboard has left the terminal. Be on it or be under it.[60]

These arguments flew with smaller broadcasters like a lead balloon. Edward P. De La Hunt, owner of De La Hunt Broadcasting in Park Rapids, Minnesota, declared that he would be "damned if any of my facilities will ever sign on to creating interference to other broadcasters."[61] Several station owners expressed renewed concern that the industry was willing to accept a degradation of analog service in exchange for the promises of a relatively untested product.[62]

The lack of receivers in the marketplace exacerbated the uncertainty surrounding HD Radio's viability. By the end of 2004, aftermarket auto HD receivers cost between $500–1,000, while tabletop units sold for more than $300 each.[63] *Radio World* blamed the lack of receivers on reluctance within the consumer electronics industry to adopt the technology,[64] while Art Reis, chief engineer for Crawford Broadcasting Company's Chicago cluster, laid fault at the feet of iBiquity. "Where is the marketing support that is supposed to help launch . . . HD Radio into the public mainstream of the electronics industry? Where are the ads on radio, TV and in print? As it stands now, iBiquity isn't even on the public's radar screens. Just walk into any

Radio Shack and ask about iBiquity or HD Radio. You've got questions, they've got blank stares . . . That, folks, is a danger sign."[65]

GROWING DISJUNCTURE BETWEEN POLICY AND REALITY

In the face of increasing industry and public consternation, HD Radio's proprietors doubled down their bets. On May 18, 2005, the National Radio Systems Committee tendered a formalized HD standard, dubbed NRSC-5, to the FCC. Such proposed standards are generally comprehensive and open for review, in order to generate industry consensus around them. But the HD standard detailed only the AM-HD and FM-HD broadcast signals themselves; it included placeholders for such features as multicasting and datacasting, due to the fact that no substantive data regarding their actual functionality had yet been tendered to the committee.[66] Petitioning the FCC to establish HD Radio as the official U.S. DAB standard would also legally preclude any discussion of or action on competing technologies. The FCC opened the proposal for public comment within a month of its submission.[67]

HD proponents lined up again to cast the technology as radio's only hope to navigate a convergent media environment. The NAB boldly claimed that the development of NRSC-5 was "open, inclusive, lengthy, exhaustive and conducted under rigorous due process procedures,"[68] and implied that "the future of radio broadcasting in the U.S. rests on" the FCC's approval of the standard.[69] National Public Radio called NRSC-5 "a significant technical milestone" and urged the FCC "to approve the standard without delay."[70] NPR also expressed confidence that iBiquity would not leverage its monopoly power over the technology.[71]

The integrity of the NRSC standard-setting process was immediately called into question. Impulse Radio, a developer of DAB datacasting applications, told the FCC about pressure within the committee to achieve consensus on NRSC-5, even though it and others had significant concerns about the lack of solid technical information on which to judge the standard. Impulse accused iBiquity of blocking moves within the NRSC to open the HD standard up for use beyond its own proprietary feature-sets. "It has, by action and inaction, influenced the time line for consideration of the various parts of the standard, and has used its leverage as the dominant voice in the proceedings to gain unfair competitive and commercial advantage," wrote Impulse.[72] Others on the committee related stories of debates that went on for "months and months" as it wrestled with proposing a standard that had "missing" components.[73] Consulting engineer Barry McLarnon revealed that "a sizable number of the [NRSC DRB] subcommittee members (seven) felt that NRSC-5 was incomplete and should not be committed to a vote . . . These members were persuaded to abstain rather than cast a negative vote," shattering the illusion that true consensus existed on the proceeding.[74] None other than Microsoft argued that, by leaving significant aspects of HD Radio technology out of the standard, iBiquity could "remain outside the NRSC patent licensing policies," which raised risks that the company would exploit its licensing regime and generated "uncertainty" that was "likely to slow product development and deployment."[75]

Independent broadcasters and the public were positively apoplectic that the FCC would consider approving a standard for HD Radio without adequate documentation. Many passionately argued that the tradeoffs necessary to implement NRSC-5 were not worth the compromises it would cause to the integrity of analog broadcasting.[76] More listeners came forward with reception horror stories, including audio submissions, of digital "hissing," "hash," and "whizzing" produced by HD Radio stations.[77] They also remained suspicious of NRSC-5's proprietary leanings. Jonathan E. Hardis of Gaithersburg, Maryland, who cared enough about the issue to request membership on the NRSC, accused the body of "cross[ing] over from making technical judgments . . . to making regulatory judgments," for which the FCC was the only proper forum.[78] Hardis blamed iBiquity for defaulting on its obligation to disclose information "that they themselves had volunteered to provide" earlier in the standard-setting process.[79]

Although media reform organizations had finally begun to engage in debate over the fundamental detriments of HD Radio, by this point in time it was far too late to change the adoptive trajectory. J. H. Snider of the New America Foundation was the sole voice in the public interest community to recognize HD for what it was. "The radio broadcasters' political genius was to redefine the meaning of the word 'channel' and to develop a standard that would abide by that definition," he explained. "With this Orwellian verbal magic, they could have their cake and eat it too; they could double their spectrum holdings to facilitate their digital radio transition without calling the doubling a 'second channel.' "[80] Incumbent commercial and public broadcasters, Snider argued, had nearly a decade to "create facts on the ground" in order to foster the imposition of HD Radio with no independent oversight.[81] He also blamed the public interest community for strategic missteps that resulted from a lack of diligence regarding radio's digital future:

> During the period in the late 1990s and early 2000s when the key digital radio decisions were being made, the public interest community, and the press they educated, were focused on the low-power FM debate. LPFM only required a tiny fraction of the FM spectrum whereas [HD Radio] used up huge amounts of it. But low-power FM was nevertheless a great issue for the grassroots-driven public interest community because everyone understood FM, many individuals and organizations throughout America wanted to be their own FM broadcasters, and the time horizon for implementing LPFM suggested the closest thing you can get to immediate gratification in a spectrum policy proceeding. In the end, [HD Radio] would get more than 95% of the white space

between the FM channels but virtually no one in the public interest community would link the issues and alert the press.[82]

Snider concluded that it was "essential that the FCC stop relying on standards developed by the broadcast industry. These clever and politically motivated standards are designed to constrain, in a highly biased fashion, the range of policy options available to the FCC."[83]

The confidence expressed by HD Radio's proponents in policy proceedings was belied by an increasing sense of unease in the trade press that the technology would not find traction. More independent broadcasters related to *Radio World* their experiences with HD-induced interference, described as sounding like "garbage," "1,000 demonic cicadas," "nasty hiss," "havoc," and "digital crap."[84] Early adopters of HD Radio reported that the digital signal was not as robust as advertised, nor did it sound better than its analog equivalent.[85] Those who backed HD claimed such perspectives had no merit. "The shouting is over," proclaimed the NRSC's Milford Smith. "Let's work together on moving forward rather than lusting after an analog past which is already becoming, by technology standards, ancient history."[86]

To that end, a cartel of a dozen investor-broadcasters formed the HD Radio Alliance in 2006 to coordinate marketing, promotion, and multicast activities in major markets.[87] Alliance members committed to setting aside $200 million worth of airtime to run spots promoting HD Radio.[88] The alliance would also oversee the rollout of commercial multicast stations in such a way as to preclude digital program competition in any given market.[89] Most commercial multicast streams would initially be derivations of an existing FM station's primary format.[90] Peter Ferrara, a former executive at Clear Channel, was tapped to head the alliance: "From an anti-competitive standpoint, if the industry didn't work together, in a cohesive manner, with consistent messaging, and providing the consumer the benefit of new and unique choices, diverse choices, the technology either will take a long time to emerge or may not happen at all," he told *Radio World*.[91] But he was not unabashedly optimistic: "Since I started this job . . . I've felt somewhat like the guy in the Ed Sullivan show keeping all of the plates spinning in the air on little sticks."[92] Clear Channel also established a "Format Lab," similar to NPR's foray into multicast syndication, to provide its stations with preproduced content for secondary digital FM channels.[93] In 2007, the alliance increased its airtime-commitment for HD promotion to $250 million.[94]

Yet many broadcasters were still not buying the hype. By 2006, more than a third of Corporation for Public Broadcasting member-stations expressed no desire to "upgrade" their facilities to HD.[95] Cox Radio, Inc., an iBiquity investor, installed AM-HD on "three or four" of its stations and then turned it off when listeners complained about the interference they caused to analog broadcasts.[96] Open hostility toward iBiquity was now a regular theme in the trade press. "Why has the FCC allowed iBiquity to rape the small-market stations with exorbitant license fees, not to mention huge capital investment with little or no return?" asked Tom Andrews, president of Lake Cities Broadcasting Corp. in Angola, Indiana.[97] Jerry Arnold, the director of engineering for four stations in Terre Haute, worried that iBiquity would petition the FCC to mandate HD compatibility as soon as it realized "that a huge majority of . . . broadcasters are resisting their suggestion to go [digital]."[98]

The dearth of digital radio receivers became an increasingly sensitive topic within the radio industry. Commercial and noncommercial broadcasters reported that when they went out to their local electronics retailers and asked for "HD Radio," sales staff steered them toward satellite radio receivers; if a store had an HD display unit, more often than not it wasn't even plugged in.[99] An NPR plan to distribute 50,000 HD Radio receivers had to be scaled back to a paltry 500 because the "radios failed to materialize." In addition, since NPR did not receive permission from iBiquity to sell the receivers, it was forced instead to "share them with board members, university officials and backers of their digital conversion campaigns."[100] At the 2006 NAB annual convention, only a handful of demo HD receivers could be found on the exhibition floor.[101] Portable HD radios were nonexistent, since iBiquity's receiver chipset was still too large and power-hungry to fit into such devices.[102]

Listener demand for HD Radio remained insubstantial. Gartner Research predicted that just 9% of U.S. households would have digital radio receivers by 2009.[103] Bridge Ratings surveys of radio listenership conducted in 2006 estimated the national HD listening audience at 450,000 per week, and forecast the 2010 HD audience at just 8.84 million, "trailing well behind satellite radio and even audio streaming to mobile phones."[104] It also reported that only 13% of listeners even knew what HD Radio was, while just 7% "said they were 'very' or 'somewhat interested' in owning an HD receiver."[105] *Radio World* reported that "significant consumer resistance" existed due to suspicions "related to the benefits of HD Radio."[106] Even iBiquity's Struble revised his projections, suggesting that HD would not reach critical mass in the marketplace until 2018.[107]

By the end of 2006, *Radio World* had joined the call of many broadcasters for the FCC to consider alternative digital radio systems.[108] Others counseled a more radical course of action—dropping digitalization entirely and getting back to the fundamentals of radio's pre-1996 business model. This sentiment was best exemplified in a *Radio World* feature interview with Edward De La Hunt. His son, Edward De La Hunt, Jr., was the associate chief of the FCC's Audio Services Division and would later become its Deputy Chief of Engineering. Hunt the Younger's career with the FCC spanned the entire developmental and policymaking process of HD Radio; in 2006 he retired from the agency to work in the family broadcast business. That same year, Hunt the Elder was inducted into Minnesota's Broadcaster Hall of Fame. "Everybody is forgetting about the small markets, where profit margins, where they exist, are very narrow," said Hunt Sr. "It seems foolish to me to buy any kind of [digital] system that obsoletes what we already have . . . [Other] small owners think it's ridiculous." However, if you mentioned HD to Hunt Jr., "his eyes light up; he thinks it's the greatest thing since sliced bread. The rest of the family says 'No, no, no.' It's an FCC 'thing.' . . . They don't look at the solid technical stuff. They look at who they like and don't like."[109] Hunt the Elder then repudiated the most significant "accomplishment" of his son's regulatory career:

> [HD Radio is] not going to make my fundraising any better, it's not going to make my community service any better. I don't own these radio stations. I own the equipment. I'm a franchise holder of a license to serve the people of this country. Broadcasters need to come back to the idea that they're here to serve. If they don't want to come back to that, I guess they deserve what they get.[110]

Hunt's interview resonated with many other broadcasters. "He's right on the money and I salute his core values," responded Harvey Twite, general manager of KEDU-LP in Ruidoso, New Mexico. "What we need now is a scientist who can clone him."[111]

BLISSFUL IGNORANCE AT THE FCC

In 2007, after nearly five more years of contentious policy discourse, the FCC's Media Bureau adopted further rules governing the proliferation of HD Radio. Although the agency was not prepared to formally endorse the NRSC-5 standard, "Radio stations and equipment manufacturers need to move forward with the DAB conversion, and we need not wait until after final action is taken on the [standard] to provide such guidance to them."[112] This "guidance," among other things, gave permission for FM-HD stations to engage in unrestricted multicasting and datacasting and unleashed full-time AM-HD broadcasting.[113] Broadcasters were also "encouraged" to experiment with the all-digital mode of HD Radio.[114] Noting that iBiquity had "abided by the Commission's patent policy up to this point in the DAB conversion process," the FCC refused to regulate the company's license fee system, reemphasizing the voluntary nature of the digital transition.[115] It inexplicably claimed that "no technical support" existed to address the controversy of HD-related interference,[116] and dismissed the notion of an analog/digital transition deadline: "There is no evidence in the record that marketplace forces cannot propel the DAB conversion forward, and effective markets tend to provide better solutions than regulatory schemes."[117]

Almost immediately, members of the public filed two Petitions for Reconsideration of the ruling. Jonathan Hardis castigated the FCC for ignoring the wholly proprietary nature of the HD Radio system. "The Commission cannot hand out patents that provide better deals than the ones earned at the Patent Office . . . They stifle competition and innovation . . . The history of this proceeding is tarnished by deceit," he wrote.[118] HD Radio was "not about a cute little digital add-on to traditional radio, something that consumers may accept or reject at their option. Instead, this proceeding is about the permanent redefinition of the broadcast radio service in the United States," and should be treated with the appropriate gravitas.[119] Relatedly, the New America Foundation, Prometheus Radio Project, Benton Foundation, Common Cause, Center for Digital Democracy, Center for Governmental Studies, and Free Press disagreed with the commission's cavalier attitude regarding the fattening of every radio station's spectral footprint. "The [FCC's approval of further rules] is premised on the unexamined and unsupported assumption that the Commission is not assigning new spectrum for mutually exclusive commercial uses to incumbent licensees . . . This spectrum may be worth billions of dollars, and may allow incumbents to provide additional program streams, engage in datacasting, and provide other types of services. Yet, the FCC neither requires licensees to pay for the use of this additional spectrum nor to provide any additional benefits to the public in return for its use."[120]

iBiquity claimed both Petitions for Reconsideration provided no "new evidence or legal justification to make changes" to the commission's newly minted rules, and assured regulators that HD Radio's feature-set would provide plenty of opportunities to serve the public interest.[121] iBiquity characterized Jonathan Hardis's petition as "nothing more than bitterness for industry rejection of his views."[122] The National Association of Broadcasters asserted that both petitions were "bottomed on a faulty premise" and represented a "simply wrong" understanding of HD Radio technology.[123] National Public Radio agreed and argued that further discussion of the petitions would "harm . . . the DAB transition and the new public services NPR and others are developing" because they challenged "the basic decision to authorize the iBiquity system and not merely some incidental aspect of it. Accordingly . . . we urge the Commission to consider the potentially catastrophic consequences for the future of digital radio and the public."[124]

Hardis and his newfound public-interest allies would not be so cavalierly dismissed. Hardis asserted that iBiquity was "so bereft of counter-argument that their opposition is reduced to distortion, falsehood, irrelevancy, and ultimately, impugning the motives of those who dare to speak up."[125] The New America Foundation and others noted, "Oppositions' arguments do not obscure that the [HD Radio] system increases the bandwidth occupancy of broadcasters. At best [they] highlight the glaring need for the FCC to perform a 'reasoned analysis' and provide a clear rationale for [its] decision."[126]

The FCC's further endorsement of HD Radio encouraged its proprietors to push forward. iBiquity signed an agreement with the Rupert Murdoch-owned NDS Group to develop a pay-radio application known as conditional access. This technology, branded "RadioGuard," would allow stations to encrypt a portion of their digital audio streams and charge individual listeners for the right to access them. However, this functionality would not be backward compatible with existing HD Radio receivers.[127] The Corporation for Public Broadcasting, having already doled out nearly $30 million in HD-related subsidies by 2007, opened another funding window for digital conversion station assistance.[128] The HD Radio Alliance re-chartered itself for a third year and reported its member-stations would devote $230 million worth of airtime to HD promotion activities in 2008.[129] In May of that year, iBiquity announced it had raised another $15 million in venture capital.[130]

By this point, however, intra-industry dialogue as documented in the trade press was overwhelmingly pessimistic about HD Radio's future. Those who still supported the technology sounded increasingly strident. John Schneider, a sales manager at transmitter manufacturer Broadcast Electronics, blamed critics for slowing the HD transition: "These naysayers from within our own industry would probably complain about the quality of the lifeboats on a sinking ship . . . If we all don't get behind [HD] as our best chance for survival, the train will leave the station and we will all be looking for a new line of work."[131] Stephen Poole, the chief engineer of Crawford Broadcasting Co.'s Birmingham, Alabama, station cluster, declared that "until the naysayers can propose a real, feasible, doable and cost-effective alternative to help [radio] grow and survive in an increasingly competitive (and increasingly digital!) marketplace, I'm finding myself increasingly uninterested in what they have to say."[132] Crawford Broadcasting's Art Reis continued the trope: "[W]hy are the naysayers griping about the 'slow pace' of HD Radio acceptance? . . . It's about time that you all became part of the solution instead of part of the problem."[133]

These sentiments only served to exacerbate concerns about HD Radio's viability. More listeners submitted reports to *Radio World* detailing their HD-related interference problems.[134] In Brooklyn, New York, Steven Daniel reported that AM-HD interference at night had "made a dramatic, even shocking, difference" to his listening experience:

> I used to listen to a lot of stations in the Northeast corridor, but now I'm limited to a few local powerhouse stations . . . The rest of the dial is a useless wall of interference. I'm willing to complain about this and write a few letters, but I doubt it will do any good. I expect that, in a month or so, I'll miss listening to my radio at night and will break down and purchase a stand-alone Internet radio. I figure that's the next best thing to "real" radio.[135]

"I can testify that my coverage has indeed suffered," wrote Larry Langford, the owner of two small AM stations in Michigan. "[HD interference] has taken out an entire market for me . . . Those who support [HD] say that iBiquity will work out the kinks and have a solution. Who are they kidding?"[136] Langford called FM-HD technology "junk science," and AM-HD "science fiction."[137]

On October 1, 2007—just four months after the FCC's promulgation of further rules to promote HD Radio's uptake—16 AM-HD stations owned by iBiquity broadcast-investor Citadel Radio abandoned their digital broadcasts. *Radio World* reported the move was due to "interference complaints from listeners and stations on adjacent channels, the latter from both Citadel- and non-Citadel-owned stations in and outside the markets . . . Listeners who have complained say they hear hiss and adjacent-channel stations say they hear noise on the channel."[138] The following month, station WYSL-AM in Avon, New York, filed the first formal complaint with the FCC seeking AM-HD interference remediation. Based on "more than 100 hours" of mobile interference recordings covering "more than 700 miles in field tests," the complaint alleged that 50,000-watt CBS-owned WBZ in Boston, which neighbored WYSL on the dial, was stomping on its signal. Neither CBS nor the FCC would comment on the complaint.[139] In letters to *Radio World*, WYSL owner Robert Savage argued that the interference "poses a threat to public safety and communications . . . If CBS continues to delay and dither, we'll have no choice but to file a federal lawsuit and seek an injunction, and at the same time we may petition the FCC for a refund of our regulatory license fees. We pay good money for the use of our frequency, and WYSL is being deprived . . . by the very agency which charges for it."[140]

HD interference problems were not confined to the AM dial. Doug Vernier, retained by NPR to study FM-HD interference, reported in a front-page *Radio World Engineering Extra* article that the HD system actually allowed more digital energy onto adjacent channels than previously predicted.[141] He also debunked proponents' claims that the technology used no new spectrum:

> What we have done with the introduction of [HD] is to superimpose a new transmission method over an existing allocation system, hoping it will work. In many cases it does, but there are more cases coming to light every day where there are problems . . . There are those who say, "Don't look a gift horse in the mouth"; *the FCC gave us the use of this new spectrum,* so let's make the best of it. Being neighborly to the stations and their listeners seems to have taken a back seat over a more hedonistic view of "Let's push on and make amends for what we have done later." [emphasis added][142]

Radio World reported that several anonymous but well-respected broadcast engineers and station executives were frustrated with the "fraud" of HD Radio; they wanted to turn off the digital sidebands on their stations but were prevented from doing so by corporate management.[143] At the 2008 Consumer Electronics Show, columnist Skip Pizzi observed, "Sentiment against [HD Radio] is gradually morphing from a fringe movement to a serious threat . . . If any other major broadcast groups drop [HD], the format will be in serious trouble."[144]

By 2008, there was still virtually no listener interest in HD Radio. Without firm commitments from vehicle manufacturers to include HD receivers as standard equipment, *Radio World*'s "Guy Wire" predicted most stations would abandon their digital signals within 10 or 20 years.[145] Wilifred Cooper told the publication that he attended the fall 2007 NAB Radio Show hoping to purchase an HD Radio and was "impressed with the [HD] theme" of the convention. However, "retro radios were being sold in the NAB Radio Show store, the kind also able to play LP records. No HD Radio receivers were for sale."[146] Audience researcher Thom Moon took a trip around the Cincinnati metropolitan area looking for HD receivers and found just a handful of models in stock. Very few store clerks knew how HD Radio worked, much less how to sell its virtues.[147] Repeating the experience in New York, Moon discovered, "HD Radios were a bit easier to find . . . But education of salespeople about HD Radio technology and the ability of potential customers to sample real HD-R broadcasts are problematic."[148]

A Bridge Ratings survey published in the summer of 2007 projected "booming expected growth for Internet radio . . . and wireless Internet . . . in the coming decade, with terrestrial radio . . . essentially flat, then eroding toward 2020, and HD Radio use bubbling along only near the bottom." The number of surveyed listeners who were "very interested" in purchasing an HD receiver actually declined—from 9% to 7%.[149] Research firm Parks Associates projected HD Radio's national audience would grow from a paltry 4.2 million in 2008 to 30 million in 2012—a figure that represented just 10% of analog radio's regular listenership.[150] By mid-2008, *Radio World* declared that it had "become clear to us that radio's health and growth do not rely on any one tool or platform but rather on a willingness to be flexible, to try something new and not be afraid to fail at one project and then try another."[151] Naysayers were no longer a fringe element in the industry.

Between 2002 and 2008, HD Radio's protagonists convinced a compliant FCC to allow the technology to proliferate, regardless of the growing record illustrating its fundamental detriments. The real-world results were inversely proportional to regulatory expectations: every time protagonists "won" a policy fight with independent broadcasters, consulting engineers, and the public, tangible perceptions of the technology soured. After eight years in existence, iBiquity still did not have a viable business model. Public broadcasters saved iBiquity's bacon in more ways than one: They led the effort to develop the FM-HD multicasting feature, provided regulatory justification for HD Radio by advocating for it at critical policy junctures, and tapped into government subsidies administered through the Corporation for Public Broadcasting to provide iBiquity and its proponents with infusions of capital worth millions of dollars. The latter was especially critical, as financial support of iBiquity by commercial radio broadcasters dwindled significantly while the industry endured a recession.

HD Radio's ballyhooed promotional campaign consisted of commercials aired using unsold spot inventory on investor-stations and reached nobody of consequence, as demonstrated by miniscule listener awareness and interest. Receiver manufacturers declined to commit to meaningful production. Independent broadcasters and the listening public were well aware of and increasingly concerned with the problems of HD Radio, and they made their sentiments known clearly and articulately both to the FCC and within the trade press. This impressive swath of opposition even found belated support from many public-interest organizations that purport to represent them in the world of media policy—but they engaged with the digital radio debate long after most of its constitutive choices were made. Although the FCC unblinkingly catered to nearly every desire of iBiquity and its broadcaster-investors, no amount of regulatory permissiveness could fix a digital broadcast technology that overpromised and under delivered. As the second decade of the twenty-first century loomed, radio's digital future was murkier than ever.

NOTES

1. See Comments of Eric S. Bueneman, MM 99–325, December 17, 2002; Comments of Kevin Redding, MM 99–325, December 17, 2002; Comments of Gerry Bishop, MM 99–325, December 20, 2002; Comments of Powell E. Way III, MM 99–325, December 18, 2002; Statement of Kevin Tekel, MM 99–325, January 15, 2003, 4; and Counterproposal of John Pavlica, Jr., MM 99–325, February 11, 2004.
2. See Reply to Petition for Consideration by Scott Todd, MM 99–325, January 2, 2003; Reply Comments of Citizen from Novi, MI, MM 99–325, December 20, 2002; and Comments of Scott A. Todd, MM 99–325, May 3, 2004.
3. See "Quick Takes: Digital Radio: Gear on Order in Tampa, Tests Set in Long Beach," *Current* XXII, no. 2 (January 27, 2003), 4; and Mike Janssen, "People: Technology," *Current* XXII, no. 9 (May 12, 2003), A17.
4. Leslie Stimson, "Stations Continue To Convert," *Radio World*, June 18, 2003, 11–12.
5. Leslie Stimson, "35 Groups Embrace HD Radio," *Radio World*, February 1, 2003, 1.
6. Leslie Stimson, "Struble to IEEE: 'We're Committed to AM,'" *Radio World*, November 20, 2002, 7.
7. Leslie Stimson, "HD Radio 'Baby' Is Ready to Fly," *Radio World*, February 12, 2003, 1, 6–8.
8. "Race for IBOC? iBiquity Offers New Incentives," *Radio World*, March 1, 2003, p. 2.
9. Mike Shane, "The Real Reason for IBOC," *Radio World*, December 18, 2002, 46.
10. See Mario Hieb, "Rackley Delves Into AM IBOC," *Radio World*, March 26, 2003, 10, 12; Daniel Mansergh, "WYGY's Digital Conversion Detailed," *Radio World*, April 23, 2003, 8, 12; and John Pavlica, Jr., "Right Pew, Wrong Church," *Radio World*, May 7, 2003, 54.
11. Scott Clifton, "Take Off the Blinders," *Radio World*, May 7, 2003, 54.
12. Leslie Stimson, "HD Radio Blitz Seen For Summer," *Radio World*, May 7, 2003, 1, 10.
13. Paul J. McLane, "From the Editor: IBOC Is Here, for Better or Worse," *Radio World*, February 12, 2003, 4–5.

14. Editorial, "Those Mysterious Digital Noises," *Radio World*, January 15, 2003, 46.
15. Leslie Stimson, "Who Owns PAC?" *Radio World*, June 18, 2003, 12.
16. See Leslie Stimson, "Algorithm Concerns Slow HD Radio," *Radio World*, June 4, 2003, 1, 3; and Mike Janssen, "Upgrade Wanted for Digital Radio Sound Quality," *Current* XXI, no. 10 (June 2, 2003), A1, A18.
17. Paul McLane, "From the Editor: This Is No Time For Secrets," *Radio World*, June 18, 2003, 4.
18. Leslie Stimson, "Impact of IBOC 'Pause' Disputed," *Radio World*, June 18, 2003, 11–12.
19. See Paul McLane, "From the Editor: Walden Will Be Missed—And How," *Radio World*, August 13, 2003, 4; Leslie Stimson, "iBiquity Cuts More Employees," *Radio World*, September 1, 2003, 10; and Steve Meng, "Walden Foundation," *Radio World*, September 10, 2003, 46.
20. Leslie Stimson, "iBiquity In the Wake Of Walden," *Radio World*, August 13, 2003, 12, 16.
21. Leslie Stimson, "Will New Codec Do the Trick?" *Radio World*, September 20, 2003, 10–11.
22. Skip Pizzi, "Digital, Everywhere But Here," *Radio World*, September 1, 2003, 15–16.
23. See David C. McCork, "One Big 'Hissing' Contest," *Radio World*, September 24, 2003, 69; Edgar C. Reihl, "Disaster in The Works," *Radio World*, September 24, 2003, 69; and Jim Jenkins, "Slingin' Hash," *Radio World*, March 10, 2004, 45.
24. See David M. Sites, "DRM, a Global Standard," *Radio World*, August 13, 2003, 46; Dom Gentile, "Eureka-147 vs. IBOC," *Radio World*, October 22, 2003, 45; Philip E. Galasso, "What's My Motivation?", *Radio World*, December 17, 2003, 46.
25. Quoted in Leslie Stimson, "iBiquity Sees Finish Line in Sight," *Radio World*, November 5, 2003, 12, 14.
26. See Comments of Music Express Broadcasting Corporation, MM 99–325, June 2, 2004, 2; Comments of Infinity Broadcasting Corporation, MM 99–325, June 14, 2004, 5; Comments of Susquehanna Radio Co., MM 99–325, June 15, 2004, 2; Comments of the National Association of Broadcasters, MM 99–325, June 16, 2004, i; Comments of the Walt Disney Company and ABC, Inc., MM 99–325, June 16, 2004, 4; Comments of WHUR-FM, MM 99–325, June 16, 2004, 1; Comments of Cox Radio, Inc., MM 99–325, June 16, 2004, 2; Letter from iBiquity Digital Corporation, MM 99–325, March 4, 2005, 3; and Reply Comments of Capitol Broadcasting Company, Inc., MM 99–325, August 2, 2004, 2.
27. Comments of Greater Media, Inc., MM 99–325, June 16, 2004, 2–4; see also Comments of Clear Channel Communications, Inc. MM 99–325, June 16, 2004, 2; Comments of Entercom Communications Corp., MM 99–325, June 16, 2004, 3, 8; Comments of The Named State Broadcasters' Associations, MM 99–325, June 16, 2004, 6; Comments of Infinity Broadcasting Corporation, MM 99–325, June 16, 2004, 3; and Reply Comments of Greater Media, Inc., MM 99–325, July 30, 2004, 1–2.
28. See Comments of KUAF and WUMB Public Radio, MM 99–325, May 28, 2004; Comments of Anthony Hunt, WXEL-FM, and WMFE-TV, MM 99–325, June 1, 2004; Comments of KUAT Communications Group and John Shelton, MM 99–325, June 2, 2004; Comments of Durwood Felton, MM 99–325, June 3, 2004; Comments of WNPR, MM 99–325, June 8, 2004; Comments of Don Rinker, WILL AM/FM/TV, KUAC FM/TV, Boise State Public Radio, KRBW AM/FM, Terry Anderson, James L. Linder, Candace France, Thomas E. Richardson, and Donna L. Zuba, MM 99–325, June 10,

2004; Comments of Mark Norman, Maine Public Broadcasting, WDAV, WFIU-FM, KWGS-FM/KWTU, WVIK, KRVS-FM, WUFT/WUJF-FM, Perry Metz, and Chuck Leavens, MM 99–325, June 11, 2004; Comments of Mark Handley, Public Radio Partnership, Carolyn Day, WKMS-FM, Public Radio in Mid-America, Talbert T. Gray, James V. Paluzzi, William L. Stengel, Yellowstone Public Radio, Ms. Lee Starkel, KOSU-FM, North Dakota Public Radio, Daniel L. Campbell, Executive Committee of the Board of Trustees of Fordham University, and University Radio Foundation, Inc. (WFAE/WFHE-FM), MM 99–325, June 14, 2004; Comments of WUSF, WRTI-FM, WIUM/WIUW, WDUQ-FM, KISU Radio, Wichita Radio Reading Services, KMUW-FM, WGBH Radio, WXPR Public Radio, WKSU, WERU-FM, Darrell Penta, Roseyle C. Swig, Nancy St. Clair Finch, Vermont Public Radio, Tamara O. Breeden, KUSP and KBDH, Northwest Public Radio, WJSU-FM, Texas Public Radio, School Board of Miami-Dade County, FL, KQED, Inc., John Lilly, Public Broadcasting Atlanta, Michael K. Dugan, and WABE, MM 99–325, June 15, 2004; Comments of Ross W. Pierce, Mike Wood, Pat V. Hayes, Carol A. Cartwright, WUMB-FM, Patricia Monteith, North Texas Public Broadcasting, WHYY-FM, Curators of The University of Missouri, WJCT-FM, Oregon Public Broadcasting, KMUW Radio, WEMU-FM, WMHT-FM, WCAL, State of Wisconsin Educational Communications Board, Eric DeWeese, KSMU-FM, New York Public Radio, WCVE-FM, WAER-FM, WPRL-FM, Public Radio KUMR-FM, WVTF Public Radio, Hampton Roads Educational Telecommunciations Association, Inc., KXCV/KRNW, WBHM, KPVU, WWNO, James Madison University Board of Directors, KUNM, KEXP, KUT Radio, Aspen Public Radio, KPLU Radio, Paul Delaney, Capital Public Radio, WNCU, Station Resource Group, KCLU, Newark Public Radio, Capital Community Broadcasting (KTOO-FM), KCRW-FM, Southern California Public Radio, WDET-FM, WAMU, Minnesota Public Radio, WXPR Public Radio, WFCR-FM, WBFO-FM, Public Radio Arizona, University of South Florida, Friends of Public Radio Arizona, Eastern Public Radio, Clement Geitner, Murray State University, and Wisconsin Public Radio, MM 99–325, June 16, 2004; Comments of Garza Baldwin, J.J. Carmola, Angela Beaver Simmons, and Nebraska Public Radio Network, MM 99–325, June 17, 2004; Comments of Community Radio for Northern Colorado and WYPR-FM, MM 99–325, June 18, 2004; Comments of the WABE Board of Directors, Nevada Public Radio, and WFAE/WFHE Board of Directors, MM 99–325, June 21, 2004; Comments of Lawrence M. Kimbrough, MM 99–325, June 23, 2004; and Comments of Linda Saunders, MM 99–325; June 30, 2004.

29. See Comments of the Corporation for Public Broadcasting, MM 99–325, June 16, 2004, 3–4; and Comments of National Public Radio, MM 99–325, June 16, 2004, 3.

30. See *ex parte* Notice of National Public Radio, MM 99–325, August 13, 2004; *ex parte* Notice of National Public Radio, MM 99–325, September 28, 2004; *ex parte* Notice of National Public Radio, MM 99–325, November 23, 2004; *ex parte* Notice of National Public Radio, MM 99–325, January 19, 2005; *ex parte* Notice of National Public Radio, MM 99–325, January 24, 2005; *ex parte* Notice of National Public Radio, MM 99–325, January 27, 2005; and *ex parte* Notice of National Public Radio, MM 99–325, March 1, 2005.

31. See Reply Comments of Press Communications, LLC, MM 99–325, June 16, 2004, 5; Reply Comments of KYPK-AM, MM 99–325, August 2, 2004, 5; Reply Comments of Superior Communications, MM 99–325, August 2, 2004, 3; and Reply Comments of Timothy C. Cutforth, P.E., MM 99–325, August 2, 2004, 4.

32. See Comments of Western Inspirational Broadcasters, Inc., MM 99–325, June 15, 2004; Reply Comments of Press Communications, LLC, MM 99–325, June 16, 2004, 3; Comments of The Livingston Radio Company and Taxi Productions, Inc., MM 99–325, June 16, 2004; Reply Comments of KYPK-AM, MM 99–325, August 2, 2004, 3–4; *ex parte* Letter of the Livingston Radio Company, MM 99–325, September 8, 2004; *and ex parte* Letter to Chairman Kevin Martin from The Livingston Radio Company, MM 99–325, May 24, 2005.

33. See Letter from Clarke Broadcasting Corporation, MM 99–325, December 22, 2004; and Letter from the Livingston Radio Company, MM 99–325, October 21, 2004.

34. Comments of Reunion Broadcasting, LLC, MM 99–325, May 19, 2004, 1–2.

35. See Comments of Stephen Craig Healy, MM 99–325, June 14, 2004, 1; Comments of The Livingston Radio Company and Taxi Productions, Inc., MM 99–325, June 16, 2004, 1; Comments of William C. Walker, MM 99–325, June 16, 2004; and Comments of Arlington Broadcasting Company, MM 99–325, June 15, 2004, 3.

36. See Comments of Timothy C. Cutforth, MM 99–325, May 27, 2004, 1–2, 11; Comments of Reunion Broadcasting, Inc., MM 99–325, June 7, 2004, 2–3; Comments of WGN Continental Broadcasting Company, MM 99–325, June 7, 2004, 1–2, 4; Comments of Robert L. Foxworth, MM 99–325, June 15, 2004, 2; and Reply Comments of Reunion Broadcasting LLC, MM 99–325, July 14, 2004, 2–4.

37. Comments of David L. Hershberger, MM 99–325, June 14, 2004, 2–4.

38. See Comments of Miller Media Group, MM 99–325, May 24, 2004, 2; Comments of Radio Kings Bay, Incorporated, MM 99–325, June 14, 2004, 2–3; Comments of Edward De La Hunt, Sr., MM 99–325, June 14, 2004, 1–2; Reply Comments of Superior Communications, MM 99–325, August 2, 2004, 3; and Reply Comments of KYPK-AM, MM 99–325, August 2, 2004, 1–2.

39. Comments of Alliance for Better Campaigns, American Federation of Television and Radio Artists, Benton Foundation, Campaign Legal Center, Center for Creative Voices in Media, Center for Digital Democracy, Center for Governmental Studies, Common Cause, National Federation of Community Broadcasters, New America Foundation, Office of Communication of the United Church of Christ, Inc., Prometheus Radio Project, and Media Access Project (i.e., "Public Interest Coalition"), MM 99–325, June 16, 2004, 6.

40. See Comments of Bill Eisenhamer, MM 99–325, April 15, 2004; Comments of Kevin Redding, MM 99–325, April 29, 2004; Comments of Paul W. Smith, MM 99–325, May 10, 2004; Comments of Steven Karty, MM 99–325, June 10, 2004; Comments of Michael Erickson, MM 99–325, June 14, 2004; Comments of John Hunter, MM 99–325, June 15, 2004; Comments of Kevin M. Tekel, MM 99–325, June 14, 2004; Reply Comments of John Pavlica, Jr., MM 99–325, July 16, 2004, 2; Comments of Douglas E. Smith, MM 99–325, April 27, 2004, 3–4; Comments of Eric S. Bueneman, MM 99–325, May 17, 2004, 1, 3; Comments of Patrick Rady, MM 99–325, June 11, 2004; Comments of Gerry Bishop, MM 99–325, June 10, 2004; and Reply Comments of Edgar Reihl, MM 99–325, August 2, 2004.

41. Motion to Stay Proceedings by the Amherst Alliance, MM 99–325, June 16, 2004, 16.

42. Comments of Public Interest Coalition, MM 99–325, June 16, 2004, 67.

43. Reply Comments of iBiquity Digital Corporation, MM 99–325, July 14, 2004, 2–5.

44. Reply Comments of the National Association of Broadcasters, MM 99–325, July 14, 2004, 2, 6.
45. Reply Comments of National Public Radio, MM 99–325, August 2, 2004, 7.
46. See Leslie Stimson, "iBiquity Looks to Commercialization," *Radio World*, April 23, 2004, 16; and Leslie Stimson, "IBOC Is in Hands Of Stations," *Radio World*, May 19, 2004, 1, 12.
47. Leslie Stimson, "Clear Channel Embraces IBOC," *Radio World*, August 11, 2004, 1, 23–24.
48. See "HD Radio Planned At 2,500 Stations," *Radio World*, February 2, 2005, 2; and Leslie Stimson, "Radio Groups Speed Up Their Digital Conversion," *Radio World*, February 15, 2005, 1, 8.
49. Leslie Stimson, "Radio Groups Plan Spending for IBOC," *Radio World*, October 6, 2004, 1, 8.
50. Charles S. Fitch, "Radio's Year of Decision: 2009," *Radio World*, December 15, 2004, 14.
51. "Puerto Rico's WPRM Goes HD Radio," *Radio World*, June 8, 2005, 25.
52. "iBiquity Gets More Financing," *Radio World*, March 30, 2005, 2.
53. Daniel Mansergh, "Pioneers Share IBOC Experiences," *Radio World*, September 8, 2004, 12; Leslie Stimson, "CPB Seeks Ideas for Digital Data Services," *Radio World*, September 8, 2004, 14; and Mike Janssen, "Digital booster NPR may buy 50,000 radios," *Current* XXIV, no. 1 (January 17, 2005), 17.
54. Mike Janssen, "Rule Change A Step Toward Multichannel Tomorrow Radio," *Current* XXIV, no. 5 (March 14, 2005), 3.
55. See Mike Janssen, "Empty Multicast Channels? NPR Offers Five Fillings," *Current* XXIV, no. 8 (May 2, 2005), 1, 14, and Mike Janssen, "NPR Stocks Five Musical Streams," *Current* XXIV, no. 16 (August 29, 2005), 8.
56. Leslie Stimson, "Antenna, Power Issues Emerge for AM IBOC," *Radio World*, May 19, 2004, 14.
57. Quoted in "Can Border Worries Delay Nighttime AM IBOC?", *Radio World*, September 24, 2004, 10.
58. See Paul Shinn, "IBOC and AM Bandwidth Reduction," *Radio World*, November 17, 2004, 46; and Tom Lange, "Saint Digital," *Radio World*, December 15, 2004, 44.
59. Thomas R. Ray, III, "AM Must Roll or Risk Gathering Moss," *Radio World*, March 5, 2004, 45.
60. Guy Wire, "Guy Breaks Down the Anti-HD Radio Arguments," *Radio World Engineering Extra*, October 27, 2004, 14–15.
61. Edward P. De La Hunt, "Interference is Unacceptable," *Radio World*, July 1, 2004, 46.
62. See Vern Killion, "The DIN of IBOC," *Radio World*, July 14, 2004, 46; and Karl G. Wolfe, "The Buggy Called the Whip Black," *Radio World*, February 16, 2005, 46.
63. See Daniel Mansergh, "Digital Radio in the Real World," *Radio World*, December 15, 2004, 3, 5, and Mike Janssen, "Digital Booster NPR May Buy 50,000 Radios," *Current*, XXIV, no. 1 (January 17, 2005), 17.
64. Editorial, "Your Move, CE," *Radio World*, March 2, 2005, 46.
65. Art Reis, "HD Radio Has A Marketing Problem," *Radio World*, March 16, 2005, 37–38.
66. Report of the National Radio Systems Committee, MM 99–325, May 18, 2005.
67. Federal Communications Commission, Policy Division, *Public Notice*, MM 99–325, June 16, 2005.
68. Comments of the National Association of Broadcasters, MM 99–325, July 18, 2005, 4; see also Comments of Entercom Communications Corp., Greater Media Inc., and Infinity Broadcasting Corp., MM 99–325, July 18, 2005.

69. Reply Comments of the National Association of Broadcasters, MM 99–325, August 17, 2005, 2.
70. Comments of National Public Radio, MM 99–325, July 18, 2005, 4.
71. Reply Comments of National Public Radio, Inc., MM 99–325, August 17, 2005, 4–6.
72. Comments of Impulse Radio, MM 99–325, July 18, 2005, 2–9.
73. Reply Comments of Digital Radio Mondiale, MM 99–325, July 28, 2005, 4.
74. Reply Comments of Barry D. McLarnon, MM 99–325, August 17, 2005, 1.
75. Comments of Microsoft Corp., Broadcast Signal Lab, LLP, and Impulse Radio, MM 99–325, July 18, 2005, 5–7.
76. See Comments of Reunion Broadcasting L.L.C., MM 99–325, July 18, 2005; Comments of Robert L. Foxworth, MM 99–325, July 18, 2005; Comments of Press Communications, LLC, MM 99–325, July 19, 2005; Comments of Larry Langford, MM 99–325, July 22, 2005; Reply Comments of Jack Taylor, MM 99–325, August 15, 2005; Reply Comments of Leonard R. Kahn, MM 99–325, August 16, 2005; Reply Comments of WRPQ-AM, MM 99–325, August 17, 2005; Comments of William Barnett, MM 99–325, July 27, 2005; Comments of Richard Van Zandt, MM 99–325, July 18, 2005; Comments of Edward Jurich, MM 99–325, July 15, 2005; *ex parte* Letter from the Livingston Radio Company, MM 99–325, August 9, 2005; *ex parte* Notice from the Livingston Radio Company, MM 99–325, January 26, 2006; Comments of John Holbrook, MM 99–325, April 2, 2007; Comments of the Broadcast Company of the Americas, LLC, MM 99–325, July 18, 2005, 2–7; Comments of WRPQ-AM, MM 99–325, July 18, 2005; Comments of James D. Jenkins, WAGS-AM, MM 99–325, August 15, 2005; Comments of Holiday Broadcasting Company, MM 99–325, July 18, 2005; Statement of Timothy C. Cutforth, P.E., MM 99–325, August 26, 2005; Comments of Paul S. Lotsof, MM 99–325, January 26, 2007; Comments of Polnet Communications, Ltd., MM 99–325, February 14, 2007; *ex parte* Notice of Polnet Communications, MM 99–325, March 9, 2007; and *ex parte* Notice of Polnet Communications, MM 99–325, March 13, 2007.
77. See Comments of Doug Dingus, MM 99–325, July 14, 2005; Reply Comments of Darwin Long, MM 99–325, July 25, 2005; Comments of Eric S. Bueneman, MM 99–325, August 8, 2005; Reply Comments of Edgar C. Reihl, P.E., MM 99–325, August 15, 2005, 2–3; Reply Comments of Ronald J. Brey, MM 99–325, August 16, 2005, 1; Comments of John R. Packard, MM 99–325, August 17, 2005, 1; Reply Comments of David L. Hershberger, MM 99–325, August 17, 2005, 2–3; Comments of Edgar C. Reihl, MM 99–325, September 6, 2005, 1–2; Comments of Kevin Redding, MM 99–325, July 25, 2005; Comments of William J. Harms, MM 99–325, July 18, 2005, 1; Comments of Thomas J. Olejniczak, MM 99–325, July 12, 2005; Letter from Tom Olejniczak, MM 99–325, April 3, 2006; Comments of John Pavlica, Jr., MM 99–325, August 22, 2006, 2; Comments of Gregory O. Smith, MM 99–325, March 29, 2007, 2; Letter from David C. Schmarder, MM 99–325, October 9, 2007, 2; and Comments of James M. Wilhelm, MM 99–325, May 15, 2008.
78. Reply Comments of Jonathan E. Hardis, MM 99–325, August 17, 2005, 12.
79. Comments of Jonathan E. Hardis, MM 99–325, July 14, 2005, 1.
80. J. H. Snider, "The FCC's Imminent Radio Multicasting Vote; Will It Be Another Broadcast Industry Giveaway?" in *ex parte* Notice of the Alliance for Better Campaigns and New America Foundation, MM 99–325, July 26, 2006, 5–6.
81. Ibid., 7.
82. Ibid., 8–9.
83. Ibid., 13.

84. See Mark Carbonaro, "HD Radio Interference," *Radio World*, September 1, 2005, 37; Robert C. Savage, "A Fundamental Truth," *Radio World*, September 28, 2005, 45; Ron Schacht, "'It' Happens," *Radio World*, September 28, 2005, 45; Tom Taggart, "HD Stands for Highly Destructive," *Radio World*, October 12, 2005, 45; Jack Hannold, "Has Anyone Thought This Through?" *Radio World*, April 20, 2006, 45; Robert Hubert, "AM Should Stay Analog," *Radio World*, October 11, 2006, 46; Travis Turner, "But Does It Work?" *Radio World*, November 6, 2006, 45; Ed De La Hunt, "Leave Us Alone," *Radio World*, May 9, 2007, 46; and Paul Drake, "FCC's IBOC Report And Order," *Radio World*, July 18, 2007, 45.

85. See Robert Conrad, "Real-World Problems with HD," *Radio World*, July 19, 2006, 44; Skip Pizzi, "An HD Radio Deployment Scorecard," *Radio World*, March 14, 2007, 18–19; W. C. Alexander, "HD Radio Observations in My Truck," *Radio World*, August 3, 2005, 18–19; Dave Obergoenner, "HD Radio Delay," *Radio World*, March 1, 2006, 46; Leslie Stimson, "Recepter Gets Another Antenna," *Radio World*, May 10, 2006, 26, 28; Mike Janssen, "Radios a Missing Element in HD Radio's Slow Rollout," *Current* XXV, no. 7 (April 17, 2006), 1, 18; and Sean Ross, "First Listen: RadioShack's $99 HD Radio," *Radio World*, January 3, 2007, 37–38.

86. "Smith: 'Let's Move Forward,'" *Radio World*, May 25, 2005, 3.

87. See Leslie Stimson, "Alliance's Work Is Just Beginning," *Radio World*, January 4, 2006, 1, 6, 8; and Leslie Stimson, "Ferrara Talks the HD-R Talk," *Radio World Special Report: Multicasting*, March 2006, 3.

88. Leslie Stimson, "Alliance's Work Is Just Beginning."

89. See Leslie Stimson, "Groups Will Divvy Up HD2 Pie," *Radio World*, January 4, 2006, 1, 7; and Joe Howard, "Peter Ferrara: Launching Radio Into Digital Space," *Radio Ink*, May 8, 2006, http://www.radioink.com/listingsEntry.asp?ID=446863&PT=industryqa.

90. Leslie Stimson, "HD Radio: 'Talk It Up!'" *Radio World*, June 7, 2006, 3.

91. Leslie Stimson, "For Broadcasters, Ferrara Is HD Radio's New Point Man," *Radio World*, March 15, 2006, 21.

92. Leslie Stimson, "Ferrara Talks the HD-R Talk," *Radio World Special Report: Multicasting*, March 2006, 8.

93. Leslie Stimson, "Now the Hard Part: Programming HD2," *Radio World*, October 25, 2006, 5.

94. Leslie Stimson, "The HD-R Alliance Gears Up for 2007," *Radio World*, January 3, 2007, 3, 5.

95. Mike Janssen, "CPB Grants Don't Persuade Stations Cool to HD Radio," *Current*, XXV, no. 23 (December 16, 2006), 1, 14.

96. Alan Carter, "'If There Are Call Letters, There Is an Opinion," *Radio World*, August 15, 2007, 1, 24–25, 27.

97. Tom Andrews, "Lights Out in Smallville," *Radio World*, June 22, 2005, 46.

98. Jerry Arnold, "Skip Pizzi and HD," *Radio World*, June 6, 2007, 53.

99. See Mike Janssen, "Radios a Missing Element in HD Radio's Slow Rollout," *Current* XXV, no. 7 (April 17, 2006), 18; Leslie Stimson, "Confusion Reigns About HD Radio," *Radio World*, January 18, 2006, 3, 5; Joe Milliken, "Audio Stores, Where Are the Radios?" *Radio World*, May 10, 2006, 1, 29–30; and Holland Cooke, "Is HD Radio Dead on Arrival?" *Radio World*, March 1, 2007, 8.

100. Mike Janssen, "Radios a Missing Element in HD Radio's Slow Rollout," 18.

101. Paul McLane, "'The Talk of the Hallways,'" *Radio World*, May 24, 2006, 4–5.

102. See Skip Pizzi, "Desperately Seeking Portable HD," *Radio World*, September 27, 2006, 12, 14; and Editorial, "I Want My Portable HD," *Radio World*, January 17, 2007, 46.
103. Mike Janssen, "Radios a Missing Element in HD Radio's Slow Rollout," 18.
104. See Paul McLane, "From the Editor: HD Marketing: Time to Go 'Big-League'," *Radio World*, May 9, 2007, 4; and Janssen, "Radios a Missing Element in HD Radio's Slow Rollout," 18.
105. Mike Janssen, "Multicasts: Betting to Build HD Radio Audience," *Current* XXVI, no. 17 (September 24, 2007), A6.
106. Leslie Stimson, "HD Radio," *Radio World*, March 1, 2007, 25.
107. Anders Madsen, "HD Radio: Could It Supplant FM?" *Radio World*, January 4, 2006, 16.
108. See Scott Fybush, "DRM Proponents Showcase System," *Radio World*, June 22, 2005, 29–30; Guy Wire, "AM IBOC Tries to Get Past the Noise," *Radio World Engineering Extra*, October 19, 2005, 22–24; Guy Wire, "NAB2006 Abuzz Over IP, HD Rollout," *Radio World Engineering Extra*, June 14, 2006, 27–28; Barry McLarnon, "AM IBOC Power Levels = Mystery," *Radio World*, July 19, 2006, 5–6; H. Donald Messer, "DRM Moves Rapidly to Market," *Radio World*, September 27, 2006, 3, 8; Dave Hershberger, "A Fourth Method for Digital Audio Broadcasting?" *Radio World Engineering Extra*, December 13, 2006, 4, 14, 16; Editorial, "Rethinking AM's Future," *Radio World*, December 20, 2006, 46; Fred Lundgren, "Dump AM IBOC, Move the AM Band," *Radio World*, October 11, 2006, 45; and Larry Tighe, "AM Migration: Possible, But Not Likely," *Radio World*, February 14, 2007, 44.
109. Paul McLane, "From the Editor: De La Hunt Troubled by HD Radio," *Radio World*, February 1, 2006, 4, 20.
110. Quoted in ibid., 20.
111. Harvey Twite, "Take It From Ed," *Radio World*, March 29, 2006, 86.
112. Federal Communications Commission, Audio Division, Media Bureau, *Second Report and Order, First Order on Reconsideration and Second Further Notice of Proposed Rulemaking*, MM 99–325, May 31, 2007, 7.
113. Ibid., 8–9, 15, 19, 37–38.
114. Ibid., 10.
115. Ibid., 41.
116. Ibid., 43–45.
117. Ibid., 8.
118. Petition for Reconsideration of Jonathan E. Hardis, MM 99–325, July 9, 2007, 2.
119. Ibid., 18.
120. Petition for Reconsideration by the New America Foundation, Prometheus Radio Project, Benton Foundation, Common Cause, Center for Digital Democracy, Center for Governmental Studies, and Free Press, MM 99–325, September 14, 2007, i.
121. Opposition to Petition for Reconsideration by iBiquity Digital Corporation, MM 99–325, February 11, 2008, 3–5.
122. Ibid., 10.
123. Opposition to Petition for Reconsideration filed by the National Association of Broadcasters, MM 99–325, February 11, 2008, 3–6.
124. Opposition to Petition for Reconsideration filed by National Public Radio, Inc., MM 99–325, February 11, 2008, 8–9.
125. Reply to Opposition to Petition for Reconsideration by Jonathan E. Hardis, MM 99–325, February 19, 2008, 3.

126. Reply to Opposition to Petition for Reconsideration by the New America Foundation et al., MM 99–325, February 21, 2008, 4.
127. Skip Pizzi, "Subscription Radio Gets Real," *Radio World*, September 1, 2007, 12, 14.
128. "Newswatch," *Radio World*, November 21, 2007, 5.
129. Leslie Stimson, "HD Radio Alliance Signs Up for Year 3," *Radio World*, November 21, 2007, 14, 16.
130. "More Venture Capital for iBiquity?" *Radio World*, May 7, 2008, 24.
131. John Schneider, "Helpful Hints For HD," *Radio World*, December 5, 2007, 45.
132. Stephen Poole, "Hey, AM HD-R Critics: Got a Better Idea?" *Radio World*, December 19, 2007, 45.
133. Art Reis, "What Slow Pace?" *Radio World*, March 1, 2008, 36.
134. See Bill Sepmier, "Buzz Might Be the Problem," *Radio World*, September 1, 2007, 46; Jay Policow, "Nighttime IBOC Interference," *Radio World*, November 7, 2007, 44; Phil Boersma, "Apocalypse Now," *Radio World*, December 5, 2007, 45; Edgar Reihl, "Hear It Now: IBOC Disappoints," *Radio World*, December 5, 2007, 45; Jim Jenkins, "Badgering the 'Witness,'" *Radio World*, February 1, 2008, 36; and Bob Young, "Canaries in the Coal Mine," *Radio World*, February 1, 2008, 38.
135. Steven Daniel, "Nighttime IBOC Interference," *Radio World*, November 7, 2007, 44.
136. Larry Langford, "Let's Have a Reality Check on IBOC," *Radio World*, December 19, 2007, 5.
137. Ibid., 6.
138. Leslie Stimson, "Citadel Re-evaluates Nighttime AM HD-R," *Radio World*, October 24, 2007, 1, 23.
139. Leslie Stimson, "Rochester Station Says IBOC Interferes," *Radio World*, December 5, 2007, 1, 12, 14.
140. Bob Savage, "Find a Real Solution," *Radio World*, February 13, 2008, 46.
141. Doug Vernier, "What Are We Doing to Ourselves, Exactly?" *Radio World Engineering Extra*, December 12, 2007, 8, 10.
142. Ibid., 18.
143. Paul McLane, "From the Editor: Digital Delight or Digital Doom?" *Radio World*, September 26, 2007, 4.
144. Skip Pizzi, "Turning Some Tight Corners for IBOC," *Radio World*, February 1, 2008, 20.
145. Guy Wire, "Time of Reckoning Nears for HD Radio," *Radio World*, October 17, 2007, 24, 26.
146. Wilfred Cooper, "No HD-R Receivers At Fall Show," *Radio World*, December 5, 2007, 44.
147. Thom Moon, "Let's Go Shopping: HD Radio in Cincy," *Radio World*, March 12, 2008, 1, 10.
148. Thom Moon, "Can You Get Tuners in the Big Apple?" *Radio World*, May 7, 2008, 22, 24.
149. Paul McLane, "Study: HD Radio's 'Tepid Growth Story,'" *Radio World*, September 12, 2007, 35.
150. "Next-Gen Radio," *Radio World*, February 13, 2008, 16.
151. Editorial, "More Than Half Full," *Radio World*, May 7, 2008, 46.

6 Tweaking an Imperfect System

Even after the FCC approved a massive expansion in the latitude that broadcasters had to experiment with HD Radio, the technology was still not on solid ground. Problems of interference on the AM band effectively halted its proliferation there, and although such challenges existed to a lesser degree on FM, HD's proponents again moved in a counterintuitive direction, hoping to facilitate uptake of the technology by effectively raising the probability of harm it would cause to the legacy analog broadcast system. While technical tweaks have continued in an effort to maximize the inherent utility of the HD system, they have had little to no effect on broadcaster or listener interest in the technology.

PROPOSING A BOOST TO FM-HD POWER

In June of 2008—almost a year to the day after the FCC promulgated further rules on the operational flexibility of HD Radio—several commercial and public broadcasters petitioned the agency to raise the power level of digital FM signals.[1] The group requested the FCC allow stations to increase the power of their FM-HD broadcasts by a factor of ten (from 1% to 10% of an FM station's analog transmission output).[2] They argued that this increase was necessary because FM-HD signals could not replicate a station's analog coverage area. In fact, the digital signals were so weak that their reception was nearly impossible within buildings.[3] This flew in the face of many of the initial assertions about HD Radio that were used to sell the technology to the FCC. Unbeknownst to the rest of the industry and the listening public, the broadcasters petitioning for the power increase had already received permission from regulators to conduct field experiments on a CBS-owned station in California. According to their filing, FM-HD signals could be received "at only one of the test locations and analog reception was described as noisy or poor at 75% of the test locations." When the power was increased, the improvement was imperfect, but "the digital signal could be received reliably at 75% of the test locations and at selected locations in the remaining 25% of the buildings."[4]

National Public Radio conducted a 50-station simulation-analysis of the effects of a blanket FM-HD power increase on analog radio reception, and the results were personally presented to FCC staff. NPR's report concluded that a power hike would increase the reliable service area of the digital signal, especially in a mobile listening environment. However, it would come with a significant cost:

> Mobile analog FM covered population would be reduced an average of 26% for the sample stations. Interference would affect some stations severely in portions of their analog mobile service area: 41% could lose one-third or more of their covered population and 18% would lose more than half of their population . . . Analog FM indoor and portable covered population totals are reduced by 22% and 6%, respectively. Interference would affect some stations severely in portions of their analog indoor service area: 27% could lose one-third or more of their covered population and 16% could lose more than half of their population.[5]

Not only did the proposed increase in digital power stand to exacerbate the potential for interference between FM stations, but raising the power of FM-HD sidebands might also cause detrimental self-interference between the analog and digital components of an FM station.[6] NPR noted that while listener reports of interference under existing FM-HD power limits were "minimal," it chalked that up to public ignorance about the technology.[7]

Although NPR's analysis predicted that a tenfold increase in FM-HD power would "cause substantial interference" to analog radio listening,[8] "The question of getting from here to there without substantial penalties to analog coverage is likely a matter of successive, calculated strategies, potentially trading off some increments in analog interference risk for more digital coverage, commensurate with increases in digital receiver penetration."[9] Additionally, NPR cautioned that raising the FM-HD power level tenfold "could [double] original HD transmission cost projections"—an extremely unattractive option for cash-strapped noncommercial radio stations.[10] The tacit signal to the FCC was that NPR was willing to support some form of FM-HD power increase but worried about its implementation in uncontrolled circumstances and the implications it might have on analog listenership.

In response to this study, iBiquity Digital Corporation conducted its own tests on the feasibility of increasing FM-HD power levels. Its conclusions could not have been more different from NPR's, asserting that such operation dramatically increased the robustness of digital signals without any "meaningful increase in the potential for harmful interference,"[11] and suggested that industry consensus on the topic had already been achieved: "broadcasters have become comfortable that the introduction of digital broadcasting does not present a credible risk of harmful interference to existing analog broadcasting."[12]

The National Association of Broadcasters, CBS Radio, and the HD Digital Radio Alliance all expressed support for a blanket tenfold power increase. Although the NAB acknowledged that the proposal "may create new instances of interference in certain situations," it believed "that the benefits to be gained for FM broadcasters and FM listeners will far outweigh the limited additional interference predicted by iBiquity's studies."[13] This was an argumentative point straight from the script to promote the imposition of full-time AM-HD broadcasting. iBiquity executives met with senior staff from the FCC's Media Bureau in October 2008 and made a PowerPoint presentation on the need for a blanket FM-HD power increase. The presentation spent much of its time downplaying the negative findings of NPR's interference study and pledged cooperation between all segments of the radio industry toward the goal of raising the allowable digital power.[14]

On October 23, 2008, the FCC's Media Bureau issued a public notice formally requesting comments on the proposal.[15] iBiquity hammered home the notion that a power increase was necessary to make FM-HD reception viable, especially for its multicasting application which, unlike a station's primary programming, had no analog fallback.[16] The company also cast NPR's interference analysis as unrealistic. "Throughout its many years of testing the HD Radio system, iBiquity has always found field tests offer the most accurate prediction of digital compatibility and performance," it claimed, and suggested NPR's report painted a "worst case scenario" regarding the implications of a power hike. The company also expected that "digital receiver penetration levels will be much more dominant long before all stations have adopted digital broadcasts. At that point, the impact of digital on existing analog signals will become much less important."[17] This, of course, assumed that many more stations would adopt the HD system.

The "Backyard Broadcasting Group," a benign moniker for a plethora of iBiquity investor-companies such as CBS, Bonneville, Clear Channel, Emmis, Entercom, Greater Media, and several transmitter manufacturers, filed comments encouraging expeditious action on the issue. It suggested the power increase would have "a tolerable potential for interference" in "limited" situations.[18] Noting that NPR had "publicly stated that it is not fundamentally opposed to—and sees the need for—a digital power increase," the group similarly painted NPR's interference analysis as an "unrealistic prediction" of any real-world impact such a change might have.[19] It further reminded the FCC that the industry had collectively spent "in excess of $225 million on HD Radio technology" to date, and that a denial of this request would put that investment in jeopardy.[20]

The majority of public broadcasters reacted negatively to iBiquity's assertion that any power increase would be painless. Minnesota Public Radio, a signatory to the initial request, formally dissociated itself with the originating petition.[21] "We are . . . committed to the integrity of the FM spectrum as a whole, and it would be inconsistent for us to . . . object to the many incursions on the integrity of the spectrum . . . while at the same time ignoring the interference that could be caused by an indiscriminate increase in power for

HD," it wrote. "MPR recommends that the Commission seriously consider approving a solution that allows the broadcasters to find an HD power level that works with their neighbors and would not sacrifice analog reception."[22] The Association of Public Radio Engineers observed that both the iBiquity and NPR studies projected a "significant" increase in digital-to-analog interference and requested the FCC explore alternate methods to help improve the robustness of FM-HD signals.[23] One suggestion would allow stations to asymmetrically increase the power of their digital sidebands: if an FM station had a nearby neighbor on one side of its channel but not the other, perhaps that station could increase its digital power in the spectral direction where interference was less likely to present a problem.[24]

NPR strongly defended the validity of its interference analysis, detailing the care and transparency with which it had accumulated its data.[25] Conceding in principle that an increase in FM-HD power was necessary, NPR asserted that a blanket tenfold hike would be "insidious" and could endanger the fiscal viability of public radio stations.[26] It announced that it "had begun intensive technical planning to carry out additional testing to address this and related issues in the next few months," and called on the FCC to defer any decision until that information was assembled.[27]

Equipment manufacturers were of mixed minds about an FM-HD power increase. Only one transmitter manufacturer, Energy-Onix, filed comments, and those were firmly against a power hike. "I strongly recommend that the FCC does not permit any increase in the existing HD radio subcarrier powers," wrote company president Bernard Wise. "This action will destroy the economics and technical reception of FM broadcasters."[28] On the other hand, two automobile manufacturers—Ford and BMW—supported the power increase. Both construed the proposal as key to stimulating listener acceptance of the technology.[29] The Consumer Electronics Association was more circumspect about the implications of a power hike, suggesting that the FCC look outside iBiquity and NPR for more forthright analysis of HD interference issues.[30]

Independent broadcasters redoubled their efforts to oppose any exacerbation of HD Radio's fundamental flaws. Reising Radio Partners, the licensee of an AM/FM combo in Columbus, Indiana, told the FCC that the National Association of Broadcasters no longer honestly represented the interests of all radio licensees.[31] Broadcast engineer Robert R. Hawkins of Edinburgh, Indiana, offered an analogy he hoped regulators would easily understand: "While the current system technically does work at 1% injection, it doesn't work well. The request for an increase from 1% to 10% injection is not unlike building a 'trucks only' lane on the interstate and later being told that for it to work properly, the speed limit would have to be increased to a level that would not be safe for traffic in adjacent lanes. This should have been considered before the [HD] standard was approved."[32] Hawkins, along with other independent broadcasters, provided observations of interference caused by existing FM-HD power levels in a variety of markets.[33] Several stations used NPR's interference-prediction methodology to calculate the disruption to their analog service areas that would occur should a blanket power increase be approved.[34] All believed any increase should be conducted on a station-by-station basis and that the FCC needed to monitor the potential of increased interference very closely.[35] Small- to medium-market radio station owners—both commercial and noncommercial—openly worried that the onus of increased interference would fall disproportionately on them, thereby depriving listeners of choice in markets already deprived of program diversity.[36]

From the perspective of broadcast engineer Paul Dean Ford of Dennison, Illinois, the bottom line was, "Increased digital signals increase interference to analog signals. The FCC is mandated to provide increased broadcast service. This proposal decreases broadcast service. Gradually increasing the 'noise floor' degrades all broadcasting. The public has not requested, nor is it buying [HD Radio] . . . HD now is merely filling the spectrum with annoying noise. Do not increase the noise."[37]

Douglas Vernier, who oversaw NPR's FM-HD interference analysis project, was even more candid about its results than NPR. "It is true that U.S. broadcasters and the Commission have bought into a digital radio system that has some advantages over the current FM analog system, but it also true that it poses a threat to our analog system to which a vast majority listens," he commented.[38] Having presented numerous papers at broadcast engineering conferences and NAB trade shows on the subject of FM-HD interference,[39] Vernier thought the drive for a digital power increase was an inefficient attempt to "improve" what was a flawed technology by design. "It is not surprising that we see interference to analog FM stations since with [HD Radio] we are attempting to impose a different allocation procedure on a system for which it was never intended," he explained. "With the [HD] system we are squeezing in signals into spaces which for years have acted as guard bands between the channel assignments. Any interference that we see now will be amplified in full if the . . . power increase is given a *carte blanche*."[40]

Gaithersburg, Maryland-based Mullaney Engineering argued that degrading the listenability of analog FM broadcasting to "save" a digital broadcast technology that was "too big to fail" was unwise. "This sounds just like the arguments being offered by many financial institutions & manufacturers in their weak justification of why Congress must bail them out of the mess that in some instances they were responsible for creating. iBiquity has nobody to blame for its problems but itself."[41] It also suggested that, given the weak proliferation of HD Radio, the time was ripe for the FCC to explore other digital broadcast technologies.[42]

Members of the public also rallied to oppose this latest proposal.[43] Several expressed disbelief that broadcasters would willingly degrade the reliability of analog radio service for a digital technology that had no demonstrable listener support.[44] H. Donald Messer, who served on the National Radio

Systems Committee during the drafting of the NRSC-5 standard, reported that the NRSC had worked for "over a decade" to set the initial FM-HD power levels very carefully, with the minimization of digital interference to analog broadcasts in mind. "If the digital coverage area is not to the liking of the system's proponents . . . it is up to them to show conclusively . . . that this increase is not unduly detrimental to other authorized users of this band."[45] Jonathan Hardis argued that the lack of industry consensus around the issue was "all the more reason why the Commission should not be bullied into a hasty decision,"[46] and he savaged iBiquity's study. "The Commission must recognize [iBiquity's] test report for what it is: an advocacy document that was conceived and then edited to advance a narrow, competitive commercial interest. There is nothing inherently wrong with that—except when such a document alone might be relied upon to decide the public interest."[47]

In May 2009, the FCC issued a curious notice asking for more public comment on the idea of a power increase. It suggested that discussion about the inherent need for a power hike was unnecessary and encouraged comment on how much of an increase should be allowed.[48] This implied that regulatory approval of some digital power increase was inevitable. The NAB used the opportunity to dismiss independent broadcaster and public opposition as "without merit and contrary to the public interest."[49] "Backyard Broadcasting" touted the superiority of iBiquity's interference study and suggested that an FM-HD power increase be authorized only for commercial broadcasters, so as to give noncommercial broadcasters the additional time they desired to study the issue.[50] iBiquity proclaimed there was "a virtual consensus" among commenters that a power increase was necessary, and argued that protecting "incumbent" analog services had the potential to "choke off the emerging technology."[51]

iBiquity and Charles River Broadcasting Company—a subsidiary of broadcast-investor Greater Media—filed additional "technical studies" in support of a tenfold FM-HD power hike. "Put . . . bluntly, in many building types, digital reception is simply impossible on well performing table model receivers and similarly nonexistent on the new class of portable receivers about to be introduced into the market place . . . Although an incremental digital power [increase] may serve to partially mitigate the situation . . . only a full 10 dB increase will permit reliable service to portable receivers and result in a close approximation of analogue coverage, two very basic and critical listener expectations."[52]

National Public Radio stood by its interference analysis and opposed any "interim" FM-HD power enhancement. It reported that it had convened a "Peer Review Group" of "other industry parties" to look for a way to implement a managed FM-HD power increase.[53] This was "no small matter," argued NPR: "Unless the Commission is prepared to revisit the evolutionary approach of transitioning from analog to hybrid analog/digital and, eventually, to a purely digital radio system, the Commission cannot grant the proposed power increase and simply disregard the adverse consequences for analog service."[54] To the extent that commercial supporters approved of a blanket power increase, NPR argued, "they largely rely on diversionary rhetoric rather than addressing the technical merits of the issue."[55]

Several public broadcasters supported this cautious approach, suggesting that any further experimentation should be transparent, collaborative, and incremental.[56] They had good cause to be skeptical. Rhode Island Public Radio, the licensee of WRNI-FM, reported to the FCC that iBiquity's earlier field experiments with increased FM-HD power levels had caused destructive interference to its analog signal. They contacted technical experts at NPR to help document the problem. As they did so:

> Greater Media and iBiquity came on the scene and insinuated themselves into the testing and evaluation of the reception conditions in all of the measurements. [A Greater Media representative] was present in the NPR van, which was tailed by an iBiquity van occupied by iBiquity and Greater Media personnel . . . It was apparent . . . that the purpose of Greater Media's and iBiquity's presence was not to promote accurate, real-world results, but instead to minimize any evidence of actual interference to the analog signal of [WRNI].[57]

Greater Media called this accusation "egregiously distorted or fallacious," based on "allegations and innuendo that have no factual underpinning" that would "undermine the integrity of the FCC's procedures."[58] iBiquity asserted that the "caustic nature" of RIPR's interference report not only harmed the development of industry consensus on the issue, but cast significant doubt on the objectivity of NPR's own interference analysis project. It challenged NPR to "make its personnel available to the Commission staff to address any concerns the Commission may have," and suggested that NPR's "continued silence on this subject will ultimately undercut the industry's and the Commission's confidence that the test program was conducted in an appropriate fashion."[59] Unable to impugn the integrity of NPR's findings directly, HD proponents instead settled on smearing the integrity of NPR itself.

Consulting engineers warned the FCC it was treading into uncharted territory without adequate scientific data to make a decision on the issue. Barry McLarnon called it "remarkable" and "foolish in the extreme" that the commission would even consider a power increase, given the lack of respectable technical justification. He characterized iBiquity et al.'s FM-HD experiments as "subterfuge" and demonstrative of a "lack of critical analysis that has been a characteristic of the [HD] system since its early days."[60] The power-hike proposal, concluded McLarnon, "ignor[ed] the laws of physics."[61] Referring to "numerous situations where station engineers and management have noted [HD] interference," Douglas Vernier cautioned the FCC not to make new policy just to "see how many complaints we get."[62] Klein Broadcast Engineering worried that "[t]he integrity of the analog FM Broadcasting Service is on the line here,"[63] and suggested that "a significant number of instances" of FM-HD-related interference already existed.[64] Some independent broadcasters, cognizant of the fact that the FCC seemed predisposed to sanction some sort of power increase, demanded the agency implement a proactive remediation program to address any increase in digital-to-analog interference that might occur.[65]

Radio listeners reiterated the fact that they simply did not want HD Radio. Gregory Smith characterized the technology as "all about what broadcasters want, not consumers . . . Few HD Radio . . . receivers have been sold, and most have been returned for dropouts, poor coverage, and bland programming."[66] James Wilhelm believed the success or failure of HD Radio should rest in the hands of receiver manufacturers and listeners, "not on a handout from the FCC . . . I challenge iBiquity to prove the need for a power increase based on solid evidence of consumer acceptance."[67] He noted, "The general public is silent and slow in understanding the scope of the transition to digital radio giving commercial interests a louder voice. Is it the desire of the Commission to take advantage of this lack of public understanding and allow the interests of [HD proponents] to take precedence by essentially force-feeding a hybrid system on the public?"[68] The Prometheus Radio Project echoed these concerns to argue that the preservation of analog FM broadcasting should remain the FCC's priority for the foreseeable future.[69]

RE-IGNITING THE VIABILITY DEBATE

Trade press discussion of the proposed power increase was feisty. Among commercial broadcasters, *Radio World* reported in 2008 that many engineers felt "anxiety and even doubt about the success of HD Radio."[70] Bert Goldman, the vice president of engineering for Independence Media and a member of the NRSC, told the publication, "everyone I've turned my [FM-HD interference projections] over to has gone into hiding or said it's conservative . . . I would like nothing more than to have someone tell me that I'm all wrong and why. Unfortunately, nobody has yet done that, and if my suspicions are correct, then substantial harm could be inflicted on hundreds of FM analog stations that count on their fringe . . . signals." Goldman was also displeased that power increase proponents completely circumvented the NRSC in its testing to justify the hike to the FCC.[71] Regulators would ultimately ignore this important point—one that was critical to selling the technology for FCC approval in the first place.

Several broadcasters noted that, as of 2009, there was still no standard way to monitor an HD Radio signal, which opened up significant potential for all sorts of interference-related problems.[72] Dave Obergonner, a broadcast engineer in St. Louis, reported that he personally measured FM stations running HD sidebands in excess of their licensed *analog* power. "This is a basic, fundamental flaw in the HD Radio system that will cause all kinds of grief in the future if [digital] power levels are increased, and the radios are in greater circulation . . . The only winners in this game will be the larger group-owned, larger-market stations who can afford this. Smaller-market (mostly independent) stations will pay a dear price nonetheless, with considerably increased interference . . . The rich get richer, and the smaller stations go bankrupt."[73]

Robert Savage, whose AM station already suffered from ongoing HD-related interference, opined that it was "tragic how we continue to argue and tinker with a technology left behind by the listening public and the majority of broadcasters long ago, while wireless Internet radio and other compelling platforms thrive and grow."

> This is not the time to pit broadcaster against broadcaster over HD Radio interference issues. We need to present a united front, clean our programming house and offer real, not illusory value to our listeners and advertisers instead of endlessly wrangling over a technology nobody in the real world cares about.[74]

These sentiments were strong enough that *Radio World* published an editorial urging HD Radio's proponents and the FCC to not "blindly move forward with an across-the-board digital power increase that robs Peter to pay Paul. Legacy stations deserve protection from interference."[75]

Among public broadcasters, sentiments were similarly divided. On one hand, NPR's HD Radio point-man, Mike Starling, reported that a blanket power increase would result in problems "to the detriment of existing analog FM signals," and that the potential for interference would vary "greatly from city to city."[76] On the other, Caryn G. Mathes, the general manager of WAMU-FM in Washington, D.C., and a self-proclaimed "evangelist for HD Radio," contextualized an FM-HD power hike as necessary to promote the vitality of its multicast feature.[77] Yet public radio's infatuation with multicasting itself was waning. In 2009, NPR found itself with an "unexpected shortfall in revenue," which led the organization to deprioritize investments in HD programming and make cuts to its workforce.[78] One of NPR's multicast filler-feeds, the Classical Public Radio Network, shut down due to a lack of new funding and weak station subscription-growth.[79] A *Current* survey of public radio executives found that most stations involved in multicasting did so with canned programming. "Enthusiasm was muted" for HD Radio more generally: most believed the technology would never become a major component of terrestrial radio broadcasting; 20% of respondents expected HD Radio to wither and die on the vine.[80] Tom Ammons, a technician with public station WQED-FM in Pittsburgh, observed that HD Radio was "stalled": "With less than 1 percent of our listening audience capable of HD listening, it's hard to argue that our management should put resources into multichannel programming."[81]

Michael Baldauf, a programming consultant in Pueblo, Colorado, reiterated the argument that better content, not new technology, would entice listeners back to traditional radio broadcasting. "You can come up with a bazillion channels, digital or analog, but if it all comes out of the same pre-recorded, canned, MP3-quality, low-creativity cookie cutter, the audience is not going to be there," he warned.[82] *Radio World* editor Paul McLane questioned whether anyone in the industry even had the knowledge to "appeal in a compelling way to a modern media consumer, particularly someone born after, say, 1980."[83] "Perhaps the problems started with deregulation," suggested Bill Parris, who managed several stations in Washington, D.C., and Baltimore. "Rolling up losing stations into market clusters reduced the root of radio's resilience . . . The long-term effects are proving near-fatal: a lack of new ideas, a lack of creators, and a management structure driven exclusively by cost reduction . . . For more than a decade we have been making a lousy product and getting away with it."[84]

In 2009, HD Digital Radio Alliance president Peter Ferrara resigned from his post, citing "time for a change."[85] iBiquity also laid off "an undisclosed number of people," though CEO Robert Struble assured the industry that the company was in firm financial shape.[86] At that year's NAB Radio Show, discussions between commercial and public radio station engineers and managers over the issue of an FM-HD power increase were described as "acrimonious," with "some very unpleasant hallway talk that lent a sour 'us vs. them' feel to the proceedings."[87] *Radio World* scored an interview with FCC Audio Division chief Peter Doyle, who remarked that "it's clear to the commission that the current power level is 'fraught with problems' and . . . the agency has seen a slowdown in the number of notifications from stations" converting to HD. Although the commission had received interference complaints, no formal action was being taken on them, and the burden of proof necessary before the FCC would investigate was left undefined.[88] This was an astounding admission considering that at least two AM stations had filed formal interference complaints, and the record of the HD Radio proceeding was replete with comments detailing problems on both the AM and FM bands.[89]

In reality, a compromise on the power increase issue had already been reached between those broadcasters that mattered most in the eyes of the FCC. In November 2009, NPR filed an *ex parte* report calling for a blanket fourfold increase in the power of FM-HD sidebands, with a maximum tenfold power hike allowed, provided stations could show little risk of destructive interference. The report was the product of collaborations between NPR, iBiquity, CBS Radio, Clear Channel, and Greater Media,[90] and utilized a very small sample of stations for actual field analysis.[91]

NPR and iBiquity issued a news release touting their collaboration. "We are delighted that the radio industry is now poised to push this technology ahead together," said iBiquity CEO Struble. "We've found practical and balanced solutions that will greatly improve reception while limiting interference to existing analog operations." NPR's Mike Starling claimed public broadcasters "are optimistic about the future of HD Radio . . . and eager to continue to work with iBiquity on the developments that will make this power increase work to everyone's advantage—stations, listeners, and receiver makers."[92] The National Association of Broadcasters declared that the deal "will enable digital radio broadcasters to improve digital coverage, better replicate their analog service and insure reliable reception of new multicast signals . . . The record in this proceeding is now complete."[93] iBiquity, the NAB, and NPR followed up quickly with personal visits to FCC headquarters, urging expeditious action on their proposal.[94]

The speed with which this compromise was reached left few with adequate time to criticize it. The New Jersey Broadcasters' Association provided the most substantive critique, calling any increase in FM-HD power levels "a violation of the spirit and letter of the law, substantively adverse to public policy, and counterproductive to the rules promulgated by the FCC."[95] The power proposal worked "at cross purposes to achieving the goals both Congress and the Commission have long expressed: localism, diversity of voices, equal opportunity, and intelligent local content. The obvious truth is that this HD power increase is being pushed through the Commission without reasonable opportunity for parties adversely and irreparably impacted by this pronouncement from even being heard as to what type of 'deal' is being considered."[96]

FCC: SEE NO EVIL, HEAR NO EVIL

On January 29, 2010—less than three months after the compromise had been unveiled, and without formal solicitation of any public comment on it—the FCC blessed the NPR/iBiquity request. After observing that the number of FM stations applying for HD broadcast authorization had been in decline since 2007,[97] the FCC asserted that allowing a power increase would change this trajectory.[98] Based on "analysis and data, as well as five years of interference-free FM hybrid digital operations . . . we are convinced that an immediate voluntary [fourfold digital power] increase . . . is appropriate, with the option for stations to request a full tenfold power hike."[99] Remarkably, on the question of interference remediation, the FCC put the onus on victim-stations to provide:

> at least six reports of ongoing (rather than transitory) objectionable interference. For each report of interference, the affected FM licensee must submit a map showing the location of the reported interference and a detailed description of the nature and extent of the interference being experienced at that location . . . The complaint must also contain a complete description of the tests and equipment used to identify the alleged interference and the scope of the unsuccessful efforts to resolve the interference.[100]

Historically, those who propose to expand or repurpose the use of occupied spectrum bear the regulatory burden to justify that their activity will not harm incumbent users. This decision turned decades of spectrum integrity policy on its head. Additionally, the technical knowledge, time, and resources necessary to satisfy this burden of proof were far outside the scope of an independent broadcaster or listener; listeners themselves were cut out of the interference-reporting process completely. It sent a strong message that the FCC didn't really want to be bothered with the real-world implications of HD Radio, again promoting the technology on the basis of willful ignorance about its fundamental detriments and in line with the stringently neoliberal principles under which contemporary media regulation occurs.

The FCC's decision sent shockwaves through consulting engineers, independent broadcasters, and the public. Several immediately filed appeals of the latest order. Mullaney Engineering was aghast: "It is an insult to the existing FM licensees, some of which have paid hundreds of thousands of dollars to the Federal Treasury to secure their licenses, to pretend that [HD Radio] does not cause interference to existing Analog FM service areas because vested interests in iBiquity's . . . radio system have negotiated or redefined the definition of what is 'objectionable' interference."[101] Alan Jurison questioned the opaque way in which NPR and iBiquity achieved their compromise and the manner by which the FCC decided the issue.[102]

Press Communications accused iBiquity of "blatantly misrepresent[ing] the system they promoted,"[103] and excoriated the FCC's new interference-reporting requirements: "Listeners today have a myriad of choices that did not exist decades ago and are about to get more. To suggest that a broadcaster would have to aggregate 6 continuous complaints and then have to wait months for the complaints to be remediated is ridiculous," equivalent to "asking people to explain their sightings of UFOs."[104] Jonathan Hardis claimed the FCC "exceeded its delegated authority" by blindly accepting the joint parties' power increase compromise.[105] "The proposition of a tenfold power increase, to put it bluntly, is primarily a gambit to occupy spectrum that would otherwise be put to other uses . . . Here, the remedy . . . is vastly disproportionate to the supposed problem."[106] The Prometheus Radio Project noted that the FCC had effectively ignored its own record, where HD Radio's fundamental weaknesses had been so well documented.[107]

The National Association of Broadcasters reacted strongly to this backlash. It asserted the FCC acted properly on "the basis of a well-developed and lengthy record" and did not ignore critical comments—it "simply disagreed with the objectors' arguments."[108] iBiquity argued that opposition to a power hike was based "principally on those parties' overall displeasure with digital broadcasting rather than specific issues with the digital power increase authorized" by the commission.[109] NPR reiterated that consideration of any appeal would "upset . . . the digital radio transition itself."[110] The FCC had previously dismissed such appeals without prejudice,[111] and was encouraged to do so again.[112] Jerry Arnold, the director of engineering for broadcaster Midwest Communications, told *Radio World* that he was "dubious (and that's being really polite) regarding the FCC's ability (or even desire) to respond quickly to any interference complaint, no matter how well-documented it may be. We all have seen the wink and a nod the FCC has given to the many [AM-HD] complaints, despite similar verbiage in their edict that allowed nighttime use of [HD]," and he expected the same pattern of regulatory behavior to continue in the FM environment.[113]

Stations did not leap at the chance to implement higher FM-HD power levels, leading the FCC's Peter Doyle to term industry response to the opportunity "disappointing." A year after the passage of the latest order, only 150 FM stations had applied to boost their digital power.[114] Just 135 stations added any HD functionality in 2009.[115] By April 2011, the FCC reported that only 1,627 radio stations—16% of the national total—had installed HD Radio itself, with the number of new implementations slowing to "maybe a handful" each month.[116] That summer, only 300 FM-HD stations were operating at increased digital power levels.[117] *Radio World* editor Paul McLane lauded the "remarkable support from our nation's regulators," but cautioned, "The day has not yet come when digital FM has earned priority over analog."[118] Part of this was due to the cost of the power hike, which Greater Media vice president of engineering Milford Smith estimated could run "into six figures" per FM station, in addition to the initial costs of adopting the HD system. Yet Smith thought such investments were worth it, as the "ultimate goal" of any power increase was "to make the analog and digital service areas at least equal."[119] Although proponents worked hard to convince the FCC that HD Radio was a quantum leap forward in broadcast technology, in the trade press they still admitted that the system fell short of even analog performance standards. The degree of disjuncture between the regulatory record and what public debate existed on the issue could not be more telling.

Despite this, NPR engineering executives encouraged public radio stations to adopt a higher FM-HD power sooner rather than later—especially before any new analog stations entered the market, as "listeners will be more likely to complain about IBOC interference, which is indistinguishable from noise, if it begins after they develop expectations about the sound of an analog station." The executives also noted that the potential of digital-to-analog interference depended on the content of colliding stations: "The most vulnerable . . . are those running news/talk or lightly processed, high-fidelity music with extended passages at low volume levels . . . And monaural stations will be less susceptible than stereo stations."[120] Two FM-HD-related interference complaints have since been formally filed with the FCC, but there is no regulatory inclination to investigate them.[121] Eighteen months after the FCC's order allowing the FM-HD power boost, *Radio World* published a two-part article by Dave Hershberger, a senior scientist at transmitter-manufacturer Continental Electronics, detailing the increased likelihood of self-interference between FM analog and boosted digital signals, and urged stations to proceed with extreme caution on any power increase.[122]

Yet proponents of the technology continued to tweak the system in hopes of maximizing its potential utility. Cognizant of the fact that increased digital power does risk increased interference, iBiquity, NPR, and others engaged in a crash-testing program regarding the notion of raising FM-HD sideband power asymmetrically. iBiquity and NPR met with senior FCC staff to make the case for asymmetric sideband operation.[123] Both organizations published studies involving just two stations that showed increasing the power of just one digital sideband did improve the robustness of FM-HD signals, though the improvement was not dramatic.[124] The FCC formally sought public comment on the practice in 2011, more than a year after the first tests had been conducted.[125]

Like previous proposals, proponents lined up in support of asymmetric sideband operation and urged the FCC to approve such operation without delay.[126] Within the industry, only one commenter—transmitter manufacturer Nautel—cautioned that stations wishing to operate with asymmetric FM-HD power needed to "self certify" that such operation would not cause increased interference, both to neighboring stations and its own analog signal.[127] The lone objector to this proposal was Jonathan Hardis, who reiterated the technical disingenuousness that had pervaded the entire HD Radio rulemaking, of which this latest request was just another example.[128] iBiquity and the NAB excoriated Hardis in response, calling his skepticism an example of "Mr. Hardis' long-standing opposition to all aspects of [HD] technology rather than any valid criticism of the asymmetric sideband proposal," and urged the FCC to not let his objections stand in the way "of advancing a sound technical proposal with unanimous industry support."[129] However, in a sign that HD proponents were not completely dismissive of Hardis's concerns, they tendered a statement signed by a dozen prominent broadcast conglomerates and transmitter manufacturers disputing his allegations.[130]

To date, the FCC has not formally approved a blanket authorization for the asymmetrical operation of digital FM sidebands, though stations can broadcast in this configuration once they have received "experimental authorization" from the agency, and it has been generous in handing out such waivers. That said, the number of FM-HD stations operating in this configuration are a relative handful—much fewer than the number of stations that have increased their digital power, and those represent just a fraction of the stations that have adopted HD Radio itself. Regardless, the NRSC has incorporated asymmetrical sideband operation into its standards documents for the system and continues to press the FCC to adopt this standard as the foundation for its technical rules.[131] Furthermore, proponents are experimenting with the deployment of small standalone booster-stations that would transmit only the digital portion of an FM-HD signal in areas where a power increase has not resolved coverage issues.[132] At present, the notion of low-power supplemental digital stations has not been advanced to the FCC as a standard mechanism to further refine the FM-HD system, but considering the technology's developmental history, it would seem to only be a matter of time before it comes to the fore—and it suggests that the inherently tenuous nature of HD transmission and reception will remain a problem with the technology for the foreseeable future.

The entire campaign to boost FM-HD power should have served as a wake-up call to broadcasters, listeners, and the FCC that the technology's fundamental detriments were endemic and might never be overcome. This campaign also laid bare the rank chicanery of previous claims about HD Radio's functionality and industry support—claims on which the foundation of U.S. digital radio policy rests. Instead, and in line with all previous technical and regulatory developments involving the technology, specious "testing," coupled with bombastic rhetorical alliances more based in economics than science, carried the day. Regulators essentially indicated that they had no interest in a deeper understanding of the technology and its challenges, much less making sure that it functioned without harm to the legacy analog broadcasting system. Their inaction on issues ranging from asymmetrical FM-HD sideband operation to the approval of a formal standard for HD Radio itself provides proponents with maximum latitude to continue their experiments with little hope of public oversight. In a nutshell, the deferral of the FCC to engage in and make substantive decisions about the technical merits of HD Radio is tantamount to blanket approval of whatever its supporters desire—post-hoc endorsement is nearly assured. Despite this, continued opposition to HD Radio from independent broadcasters, engineers, and the listening public suggests that "the marketplace," which proponents and regulators assert is the ultimate arbiter of HD Radio's success or failure, seems to be leaning toward the latter.

NOTES

1. Letter from American Public Media Group et al., MM 99–325, June 10, 2008.
2. Ibid., 1.
3. Ibid., 2–3.
4. Ibid., 5–6.
5. National Public Radio, "Report to the Corporation for Public Broadcasting: Digital Radio Coverage and Interference Analysis (DRCIA) Research Project, Final Report," May 19, 2008, filed in an *ex parte* Letter and Report from National Public Radio Inc., MM 99–325, July 18, 2008, 5–6.
6. Dennis Haarsager and Mike Starling, "What Public Stations Should Consider About Upgrading HD Power," *Current*, May 17, 2010, http://www.current.org/wp-content/themes/current/archive-site/tech/tech1009hdradio.shtml.
7. National Public Radio, "Report to the Corporation for Public Broadcasting (DRCIA)," 9.
8. Ibid.
9. Ibid., 22.
10. Ibid., 41.
11. Report of iBiquity Digital Corporation, MM 99–325, June 10, 2008, 3–4, 8.
12. Ibid., 2.

13. Comments of the National Association of Broadcasters, MM 99–325, June 13, 2008, 1.
14. *Ex parte* Notice of iBiquity Digital Corporation, MM 99–325, October 16, 2008.
15. Federal Communications Commission, Media Bureau, Public Notice, "Comment Sought on Specific Issues Regarding Joint Parties' Request for FM Digital Power Increase and Associated Technical Studies," MM 99–325, October 23, 2008, DA 08–2340 (May 10, 2010).
16. Comments of iBiquity Digital Corporation, MM 99–325, December 5, 2008, 2, 4–5.
17. Ibid., 7–8.
18. Comments of Backyard Broadcasting Group et al., MM 99–325, December 5, 2008, iii.
19. Ibid., ii, 13.
20. Ibid., 3.
21. Comments of Minnesota Public Radio, MM 99–325, December 4, 2008, 3.
22. Ibid., 4–5.
23. Comments of the Association of Public Radio Engineers, Inc., MM 99–325, December 5, 2008, 2–4.
24. Ibid., 5.
25. Comments of National Public Radio, Inc., MM 99–325, December 5, 2008, 5–7.
26. Comments of National Public Radio, Inc., MM 99–325, December 5, 2008, ii.
27. Ibid., 15–16.
28. Comments of Energy-Onix Broadcast Equipment Co., Inc., MM 99–325, November 25, 2008.
29. See Comments of Ford Motor Company, MM 99–325, December 2, 2008; and Comments of BMW of North America, Inc., MM 99–325, December 4, 2008.
30. Comments of the Consumer Electronics Association, MM 99–325, December 5, 2008, 5.
31. Letter from Reising Radio Partners Incorporated, MM 99–325, July 7, 2008, 1.
32. Letter from Robert R. Hawkins, MM 99–325, July 7, 2008, 1.
33. Ibid., 1–2; see also Comments of the Mars Hill Network, MM 99–325, November 14, 2008; Comments of Paul S. Lotsof, MM 99–325, December 1, 2008, 1; Comments of Charles Keiler, MM 99–325, December 8, 2008, 5–10; Comments of the Livingston Radio Co./WHMI, MM 99–325, December 5, 2008; Comments of Steven J. Callahan, MM 99–325, January 12, 2009; Comments of Carlson Communications International, MM 99–325, November 25, 2008; Comments of Nevada City Community Broadcast Group, Inc., MM 99–325, January 12, 2009; and Comments of Daniel Houg, MM 99–325, June 15, 2009.
34. See Comments of Brown Broadcasting Service, Inc., MM 99–325, December 5, 2008; Comments of WNYC Radio, MM 99–325, December 5, 2008; Comments of Mt. Wilson Broadcasters, Inc., MM 99–325, December 4, 2008; Comments of the Moody Bible Institute of Chicago, MM 99–325, December 4, 2008; Comments of Augusta Radio Fellowship Institute, Inc., MM 99–325, December 5, 2008; Comments of Houston Christian Broadcasters, Inc., MM 99–325, December 5, 2008; Comments of the Educational Media Foundation, MM 99–325, December 5, 2008; and Reply Comments of Marshfield Broadcasting Company Inc. DBA WATD-FM, MM 99–325, January 12, 2009.
35. See Comments of WHRV and WHRO-FM, MM 99–325, November 25, 2008; Comments of Radio Training Network, Inc., MM 99–325, November 26, 2008; and Comments of Delmarva Broadcasting Company, MM 99–325, November 28, 2008.

36. See Comments of Simmons Media Group, LLC, MM 99–325, November 24, 2008; Comments of Jeff Johnson, MM 99–325, November 25, 2008; Comments of Robert M. Fiocchi, MM 99–325, November 25, 2008; Comments of WOLF Radio, Inc., MM 99–325, November 28, 2008; and Comments of Leigh Robartes, MM 99–325, November 28, 2008.
37. Comments of Paul Dean Ford, P.E., MM 99–325, December 5, 2008.
38. Comments of Douglas L. Vernier, MM 99–325, November 28, 2008, 2.
39. Ibid., 3.
40. Ibid.
41. Comments of Mullaney Engineering, Inc., MM 99–325, December 5, 2008, 1–2.
42. Ibid., 3.
43. See Comments of Brian Gregory, MM 99–325, October 23, 2008; Comments of Chris Kantack, MM 99–325, October 24, 2008; Comments of Aaron Read, MM 99–325, November 12, 2008; and Reply Comments of H. Donald Messer, MM 99–325, January 21, 2009.
44. See Comments of Robert D. Young, Jr., MM 99–325, October 27, 2008, and Comments of Tim Houser, MM 99–325, October 27, 2008.
45. Comments of H. Donald Messer, MM 99–325, November 14, 2008, 2, 8.
46. Comments of Jonathan E. Hardis, MM 99–325, November 28, 2008, 4.
47. Ibid., 7.
48. Federal Communications Commission, Media Bureau, Public Notice, "Comment Sought on Specific Issues Regarding Joint Parties Request for FM Digital Power Increase and Associated Technical Studies," May 22, 2009, 24 FCC Rcd 5818.
49. Reply Comments of the National Association of Broadcasters, MM 99–325, January 12, 2009, 3.
50. Reply Comments of Backyard Broadcasting, Inc., et al., MM 99–325, January 12, 2009, 2–4.
51. Reply Comments of iBiquity Digital Corporation, MM 99–325, January 12, 2009, 1–4.
52. Report from WKLB-FM—Charles River Broadcasting Company, a subsidiary of Greater Media, Inc., MM 99–325, July 6, 2009.
53. Comments of National Public Radio, MM 99–325, July 6, 2009, 6.
54. Reply Comments of National Public Radio, Inc., MM 99–325, July 17, 2009, 2.
55. Ibid., 4–5.
56. See Comments of the University Station Alliance, MM 99–325, June 25, 2009; Comments of American Public Media Group, MM 99–325, July 6, 2009; Comments of Western States Public Radio, Eastern Region Public Media, Public Radio in Mid America, and California Public Radio, MM 99–325, July 6, 2009; Comments of Wisconsin Public Radio, MM 99–325, June 24, 2009; Comments of Seton Hall University (WSOU-FM), MM 99–325, July 2, 2009; Comments of WUKY Public Radio, MM 99–325, July 3, 2009; Comments of American University (WAMU), MM 99–325, July 6, 2009; and Reply Comments of WFCR, MM 99–325, July 17, 2009.
57. Reply Comments of Rhode Island Public Radio, MM 99–325, July 17, 2009, 4–5.
58. *Ex parte* Letter from Greater Media, Inc., MM 99–325, August 5, 2009, 1, 6.
59. Letter from iBiquity Digital Corporation, MM 99–325, August 6, 2009, 2.
60. Comments of Barry D. McLarnon, MM 99–325, June 18, 2009, 2–4, 10.
61. Reply Comments of Barry D. McLarnon, MM 99–325, July 17, 2009, 3.
62. Comments of V-Soft Communications, LLC, MM 99–325, July 6, 2009, 2.
63. Comments of Klein Broadcast Engineering, LLC, MM 99–325, July 6, 2009, 3.
64. Ibid., 2, 6–7.

65. See Comments of Press Communications, LLC, MM 99–325, July 6, 2009, 4; and Comments of the Educational Media Foundation, MM 99–325, July 6, 2009, 5.
66. Reply Comments of Gregory Smith, MM 99–325, July 7, 2009, 2.
67. Comments of James Wilhelm, MM 99–325, June 22, 2009, 1.
68. Comments of James M. Wilhelm, MM 99–325, June 29, 2009; see also Reply Comments of James M. Wilhelm, MM 99–325, July 8, 2009.
69. Comments of the Prometheus Radio Project, MM 99–325, July 6, 2009, 5.
70. Leslie Stimson, "HD Radio Coverage Is a Power Issue," *Radio World*, May 21, 2008, 8.
71. Leslie Stimson, "HD Radio Power Boost Is on the Table," *Radio World*, August 1, 2008, 5.
72. See "Newswatch," *Radio World*, April 8, 2009, 6; Leslie Stimson, "NRSC Adopts IBOC Measurement Guide," *Radio World*, May 20, 2009, 3, 5; and Jerry Arnold, "The Cart Before the IBOC Horse," *Radio World*, August 1, 2009, 53. A standardized method of measuring FM-HD signals would not be promulgated until 2010; see Leslie Stimson, "NRSC Updates FM IBOC Measurement," *Radio World*, May 19, 2010, 3.
73. Dave Obergonner, "IBOC Interference," *Radio World*, October 7, 2009, 44.
74. Robert C. Savage, "This Is Insanity," *Radio World*, November 19, 2008, 46.
75. Editorial, "Care and Study: More on the Power Increase," *Radio World*, September 10, 2008, 70.
76. Steve Behrens, "More Power for HD Radio, More Buzz on Analog," *Current* XXVII, no. 15 (September 2, 2008), 8.
77. Caryn G. Mathes, "For WAMU: HD Radio Means Higher Yields," *Current* XXVIII, no. 17 (Sept. 8, 2009), 14.
78. Leslie Stimson, "NPR Goes into New Year Leaner," *Radio World*, January 1, 2009, 3.
79. See National Public Radio, "Tough Economy Forces NPR to Address Unexpected Shortfall in Revenue," *Current,* December 10, 2008, http://www.current.org/wp-content/themes/current/archive-site/npr/npr0823nprwest.html; and Mike Janssen, "Partners to Close Classical Net for Radio but See Chance for Growth Online," *Current*, March 24, 2008, http://www.current.org/wp-content/themes/current/archive-site/music/music0805cprn.shtml
80. Steve Behrens, "60% of HD Radio Streams Not Fully Assembled Feeds," *Current* XXVII, no. 16 (September 15, 2008), A14.
81. Tom Ammons, "Power Hike," *Radio World*, September 10, 2008, 70.
82. Michael Baldauf, "Listeners at Arm's Length," *Radio World*, September 10, 2008, 70.
83. Paul J. McLane, "From the Editor: Make The Most of This New Tool," *Radio World*, December 17, 2008, 4.
84. Bill Parris, "Radio, the State of the Ship," *Radio World*, January 14, 2009, 37; see also Jim Jenkins, "Don't Tell Me It's Not a Problem," *Radio World*, October 7, 2009, 46.
85. Leslie Stimson, "Digital News," *Radio World*, November 19, 2008, 22.
86. Leslie Stimson, "The Year of HD Radio Portables," *Radio World*, February 11, 2009, 22.
87. Paul McLane, "From the Editor: In Philly, It's All About the Power," *Radio World*, October 21, 2009, 4.
88. Leslie Stimson, "FCC Ready to Move on Power Increase?", *Radio World*, October 21, 2009, 12.
89. Leslie Stimson, "AMs Dispute IBOC Interference," *Radio World*, September 9, 2009, 3.

90. National Public Radio, "Report to the FCC on the Advanced IBOC Coverage and Compatibility Study," November 3, 2009, 1, in *ex parte* Report of National Public Radio, Inc., MM 99–325, November 4, 2009.

91. Ibid., 9–11.

92. National Public Radio and iBiquity Digital Corporation, "NPR & iBiquity Strike Deal on HD Radio Power Increase," November 5, 2010, in *ex parte* Notice of National Public Radio, Inc., MM 99–325, November 6, 2009.

93. Letter from the National Association of Broadcasters, MM 99–325, November 5, 2009, 1.

94. See *ex parte* Notice of National Association of Broadcasters, MM 99–325, November 12, 2009; *ex parte* Notice of iBiquity Digital Corporation, MM 99–325, December 1, 2009; and *ex parte* Notice of National Public Radio, Inc., MM 99–325, January 28, 2010.

95. Letter from the New Jersey Broadcasters Association to FCC Chairman Julius Genachowski, MM 99–325, January 26, 2010, 1.

96. Ibid., 7.

97. Federal Communications Commission, Media Bureau, Public Order, *Digital Audio Broadcasting Systems and Their Impact on the Terrestrial Radio Broadcast Service*, DA 10–208, MM 99–325, January 29, 2010, 6.

98. Ibid., 2.

99. Ibid., 7.

100. Ibid., 10–12.

101. Petition for Reconsideration of Mullaney Engineering, Inc., MM 99–325, May 10, 2010, 4; see also Reply to Opposition by Mullaney Engineering, Inc., MM 99–325, June 10, 2010.

102. Petition for Reconsideration filed by Alan W. Jurison, MM 99–325, February 28, 2010, 12; see also Petition for Reconsideration filed by Alan W. Jurison, MM 99–325, May 10, 2010.

103. Application for Review of Press Communications, LLC, MM 99–325, May 10, 2010, 7.

104. See ibid., 15, and Response to Opposition and Motion to Stay of Press Communications LLC, MM 99–325, June 9, 2010, 14–15.

105. Application for Review of Jonathan E. Hardis, MM 99–325, March 17, 2010, i-ii.

106. Ibid., 5–6.

107. Application for Review of the Prometheus Radio Project, MM 99–325, May 10, 2010.

108. Comments of the National Association of Broadcasters, MM 99–325, May 25, 2010.

109. See Reply Comments of iBiquity Digital Corporation, National Public Radio, and the National Association of Broadcasters, MM 99–325, June 9, 2010; and Opposition of iBiquity Digital Corporation, MM 99–325, May 25, 2010, 8.

110. Opposition to Application for Review by National Public Radio, MM 99–325, April 23, 2010, 12.

111. See Federal Communications Commission, Chief, Media Bureau, Dismissal of Application for Review, MM 99–325, April 5, 2010; and Federal Communications Commission, Chief, Media Bureau, Dismissal of Petition for Reconsideration, MM 99–325, April 5, 2010.

112. Stimson, "HD Radio on the Side," *Radio World*, November 3, 2010, 10.

113. Jerry Arnold, "Dubious on Digital," *Radio World*, May 5, 2010, 45.

114. See "FCC's Doyle: Response to HD Power Boost 'Disappointing,'" *Radio Ink*, October 6, 2010, http://www.radioink.com/Article.asp?id=1977296; and Leslie Stimson, "HD Radio on the Side," 8.

115. Jim Motavalli, "HD Radio: Is It the Auto Industry's Next Big Thing?" *Mother Nature Network* via *Forbes*, January 4, 2011, http://blogs.forbes.com/eco-nomics/2011/01/04/hd-radio-is-it-the-auto-industrys-next-big-thing/.
116. "Doyle: HD Adoption Rate 'Mixed,' " *Radio World*, April 13, 2011, http://radioworld.com/article/doyle-hd-adoption-rate-%E2%80%98mixed/23245.
117. Guy Wire, "HD Radio: Is It Worth the Effort?" *Radio World*, June 8, 2011, 18, 22.
118. Paul McLane, "From the Editor: Digital Radio Cranks Up the Juice," *Radio World*, February 10, 2010, 12; see also Leslie Stimson, "How Many HD FMs Will Raise Power?" *Radio World*, March 24, 2010, 10.
119. Paul McLane, "What's Up With HD Radio Power?" *Radio World Transmission Technology*, January 2013, 12.
120. Haarsager and Starling, "What Public Stations Should Consider About Upgrading HD Power."
121. See "KATY-FM Files FCC Petition Against KRTH HD Radio Interference," *Radio Business Report*, May 20, 2010, http://rbr.com/katy-fm-files-fcc-peti tion-against-krth-hd-radio-interference/; Leslie Stimson, "CBS Radio Disputes KATY's IBOC Interference Claim," *Radio World*, July 15, 2010, http://www.rwonline.com/article/103510; Leslie Stimson, "Second SoCal Class A IBOC Interference Complaint Surfaces," *Radio World*, May 27, 2010, http://www.radioworld.com/article/101240; Richard Wagoner, "KATY, KMLA Want FCC to Run Interference with KRTH, KOST," *Los Angeles Daily News*, May 28, 2010, http://www.dailynews.com/lalife/ci_15177216; and Leslie Stimson, "The Leslie Report," *Radio World*, August 11, 2010, 6.
122. See Dave Hershberger, "Look Before You Leap," *Radio World*, August 23, 2010, http://www.rwonline.com/article/look-before-you-leap/4804; Dave Hersh berger, "Elevated HD Power, Part II," *Radio World*, October 19, 2010, http://www.radioworld.com/article/elevated-hd-power-part-ii/4819; and Brian Bee-zley, "HD Radio Self-Noise," *Ham-radio.com*, November 2010, http://www.ham-radio.com/k6sti/hdrsn.htm.
123. See *ex parte* Notice of National Public Radio, MM 99–325, September 10, 2010; and *ex parte* Notice of iBiquity Digital Corporation, MM 99–325, October 4, 2011.
124. See iBiquity Digital Corporation, "FM HD Radio™ Field Performance with Unequal Digital Sideband Carrier Levels (Preliminary)," Revision 01.03, February 22, 2011; and Hel Kellner, John Kean, and John Holt, "PAPR and Asymmetrical Sidebands Field Results: HD Radio™ Coverage Technologies," paper presented at the NAB Engineering Conference, April 2011 (filed into FCC record as part of *ex parte* Notice of National Public Radio, MM 99–325, October 24, 2011).
125. Federal Communications Commission, Media Bureau, "Comment Sought on Request for FM Asymmetric Sideband Operation and Associated Technical Studies," DA 11–1832, MM 99–325, November 1, 2011.
126. See Comments of Beasley Broadcast Group, Broadcast Electronics, CBS Radio, Continental Electronics, Emmis Communications, Entercom Commu-nications, Greater Media, Harris Corporation, Journal Broadcast Corpora-tion, Lincoln Financial Media Company, Nautel Maine, and Radio One, MM 99–325, December 19, 2011; Comments of National Association of Broad-casters, MM 99–325, December 19, 2011; Comments of National Public Radio, MM 99–325, December 19, 2011; and Comments of iBiquity Digital Corporation, MM 99–325, December 19, 2011.
127. Comments of Nautel Maine, Inc., MM 99–325, December 16, 2011.
128. See Comments of Jonathan Hardis, MM 99–325, December 19, 2011; and Reply Comments of Jonathan Hardis, MM 99–325, January 24, 2012.

129. Reply Comments of iBiquity Digital Corporation, MM 99–325, January 24, 2012.
130. Reply Comments of "Joint Parties," MM 99–325, January 24, 2012.
131. See *ex parte* Notice of National Association of Broadcasters, MM 99–325, February 7, 2012; *ex parte* Submission of the NAB and NRSC, MM 99–325, March 7, 2012; and Michael LeClair, "NRSC Revises HD Radio Standards," *Radio World*, December 12, 2012, 3.
132. See Leslie Stimson, "Fastroad Releases Early SFN Results," *Radio World*, December 15, 2010, 7, and Leslie Stimson, "FM+HD Booster Design Under Review," *Radio World*, March 1, 2012, 1, 10.

7   HD Radio's Murky Future

Going into the second decade of the twenty-first century, the collective uptake of digital audio services *not* related to broadcasting overtook the growth of HD Radio itself.[1] Part of this was due to a recession that hit broadcasters hard, which led to cuts in the resources stations could devote to the digital broadcast transition and new, HD-exclusive programming. In simple terms, many who adopted HD did so shoddily, with numerous issues arising related to the audio quality of digital radio channels, the consistency of programming on new digital-only streams, and the impact of interference caused by HD signals. These problems gave consumer electronics manufacturers and radio listeners little incentive to invest in digital broadcasting.

This has left the radio industry with a profound sense of frustration about the viability and potential of HD Radio, coupled with a sentiment of resignation about the economic and regulatory impediments to reconsidering radio's digital transition altogether. Broadcast-proponents have adopted a multi-pronged strategy to try and resuscitate interest in HD, the majority of which relies on passive or subsidized uptake strategies and platforms independent of the "traditional" digital radio infrastructure. Further innovations in HD have been hampered by the technology's proprietary nature and seem to be limited to the purview of broadcast-investors, some of who are working at cross-purposes. The net effect is a malaise that threatens to stymie radio's eventual transition to all-digital transmission.

Although the FCC has declared the marketplace the ultimate arbiter of HD Radio's success or failure, changing the present state of affairs is likely to require additional regulatory intervention. If history is any guide, this will take the form of laissez-faire permissiveness, as opposed to meaningful engagement with the issue that takes into proper consideration the perspectives of all stakeholders affected by radio's digital transition. The longer this malaise continues, the more likely it is that the marketplace definition and popular consensus of what "radio" is will continue to shift away from those historically identified with the medium. This chapter provides a comprehensive overview of the state of play regarding HD Radio in the United States from the perspectives of broadcasters, consumer electronics manufacturers, and the listening public.

BROADCASTER ADOPTION

As of April 2013, the FCC's Consolidated Data Base System (CDBS) shows that 2,038 U.S. radio stations have commenced HD broadcasts since 2002. This includes 1,733 full-power FM stations, 299 AM stations, one LPFM station, two FM translators, and three FM boosters. There are 22,138 AM and FM broadcast stations of all classifications on the air.[2] This puts HD Radio's overall broadcast penetration rate at just 9.2% after ten years; among full-power stations, which are most likely to adopt the technology, HD's penetration rate is 13.3%.

However, these numbers do not factor into consideration stations that have abandoned the technology—just those that filed initial notification with the FCC to use it. For example, FCC records indicate that nearly 300 of the nation's 4,734 AM stations (or 6.3%) have adopted HD Radio. However, Barry McLarnon's independent database of AM-HD stations on the air reports that as of June 2013, just 197 (or 4.2%) still broadcast in HD—and of those, only 64 (1.3%) operate digitally 24 hours a day. One-third of all AM-HD adopters (more than 100 stations) have abandoned the protocol completely.[3] There is no similar count of the number of FM stations that have dropped their digital broadcasts.

The Pew Research Center's Project for Excellence in Journalism suggests that HD Radio uptake by broadcasters peaked in 2005–2006 and has been in decline since then. Pew's data, represented in Figure 7.1, also shows that more stations dropped HD functionality in 2012 than added it—the first

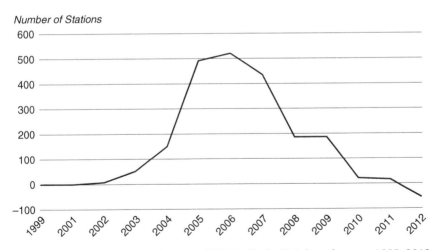

Figure 7.1   Year-over-year adoption of HD Radio by U.S. broadcasters, 1999–2012.

Source: Courtesy of the Pew Research Center, Project for Excellence in Journalism, "Number of Stations Dropping HD Outnumber Those Adopting it in 2012," in *State of the News Media 2013: An Annual Report on American Journalism*, March 18, 2013, http://stateofthemedia.org/2013/

net yearly decline since the technology's public introduction. Considering the formally unreported nature of stations abandoning digital broadcasts, the actual number of HD stations on the air is fewer than 2,000, with the majority sequestered in the top 150 U.S. radio markets.

Many FM broadcasters fail to properly maintain their HD transmissions: iBiquity Senior Vice President of Broadcast Programs and Advanced Services Joe D'Angelo told *Radio World* in 2012 that complaints of echo effects and "skipping," due to the misalignment of a station's analog and digital signals, are on the rise. In addition, "Other complaints include, 'the radio doesn't pick up HD stations, ever.' Or 'The HD goes in and out.'" These are problems directly related to the lower power of HD signals relative to their analog counterparts, and issues of interference between them.[4]

On the AM side, many major broadcast-investors in iBiquity Digital Corporation, including Bonneville, Citadel, Clear Channel, and Disney/ABC, have shut off some or all of their AM-HD signals at night due to interference concerns.[5] Although iBiquity has revised its hybrid AM-HD system in an attempt to better protect the integrity of stations' analog signals,[6] the change reduces the digital carrying capacity of the system—a move that effectively decreases its functionality.[7] Several broadcast engineers and executives contacted by *Radio World* to describe the state of AM-HD in 2010 refused to go on the record, "citing the sensitivity of the subject and company policy."[8]

Despite this sorry state of affairs on the AM band, HD proponents are actually pursuing the notion of a wholesale digital conversion. In 2011, the National Association of Broadcasters formed a task force devoted to exploring the future of AM broadcasting.[9] The task force is a project of the NAB's Radio Technology Committee, which is made up of executives from several HD broadcast-investors, including CBS, Clear Channel, Disney/ABC, Cox, Cumulus, Emmis, Entercom, Greater Media, and many others.[10] Concerned about the increasing noise floor of the band and the tenuous fiscal health of many AM broadcasters, the task force is weighing several options to improve the situation, one of which is total digitalization. In 2012, it prepared a preliminary report listing options to "revitalize" the AM band, but:

> the report is extremely technical and would need to be explained and/or possibly watered down for consumption by non-engineers. Asked why discussions about options for AM improvement are taking place behind closed doors, engineers [said] . . . they are following the wishes of the NAB board, which consists of executives from member radio groups. Closed meetings are typical in such situations, according to the leader of one NAB member company. The executive said that if such meetings were open . . . no one would feel free to participate and no work would be accomplished.[11]

According to members of the task force who spoke anonymously with *Radio World* in 2012, "Some AMs, particularly those with multi-tower directional antennas that are too narrowband to pass a digital signal, might never be able to make AM-HD Radio work without huge investments."[12] This suggests that, in an all-digital HD broadcast world, scores of AM stations may be forced to sign off because they're inherently incompatible with the technology.

Yet the task force has moved forward with all-digital AM-HD tests: in late 2012 CBS-owned WBCN in Charlotte, North Carolina, conducted several short broadcasts.[13] The NAB and iBiquity initially refused to disclose that WBCN was the guinea pig, nor would they comment on what the testing actually measured, other than to note that more experimental broadcasts on other stations are in the works.[14] According to *Radio World* reporter Leslie Stimson, "One source told me initial results look positive while another said the results really haven't been digested yet."[15] One might think that this process would be worthy of public (or at least industry-wide) deliberation, but there are no indications that the NAB ever plans to release its work.

Republican FCC Commissioner Ajit Pai may have seeded the field for this activity when he called for the creation of an "AM Revitalization Initiative" in 2012, and tasked it with producing with a plan of action for the band by as early as 2014. Thus, the trajectory for eventual fast-tracked approval of all-digital AM-HD broadcasting seems clear, and it places the ongoing "testing" regimen in proper context.[16] Although a member of the NAB's Radio Technology Committee who spoke anonymously to *Radio World* emphasized that these tests do not signal a preference for completely digitizing the AM band—"It's just the [option] we can work on first"[17]—the pattern of opacity surrounding this option is eerily similar to the behavior of HD proponents throughout the developmental history of the technology and the policymaking which has enabled it.

Considering that hybrid AM-HD technology has been dubbed "nothing more than legalized jamming" and is the bane of line-level broadcast engineers who would like nothing more than to abandon it,[18] it is difficult to wrap one's mind around the idea that AM might lead the campaign to actualize radio's all-digital transition. That said, the time to make such a radical change is fleeting: Jim Paluzzi, the general manager of public radio stations in Phoenix, Arizona, predicted in 2010 that AM-HD Radio "will have only about five years to establish a user base" before the window of opportunity for its adoption closes.[19]

Broadcast-investors in iBiquity no longer have the fiscal wherewithal to directly support the company, as reflected by an enormous devaluation in publicly traded radio company market capitalization over the last decade. Many radio stocks now trade at a fraction of their all-time highs. Clear Channel, the nation's largest radio broadcaster, re-privatized itself with the assistance of Bain Capital and other private "vulture funds" after taking a beating on Wall Street, and now carries a debt of more than $20 billion, more than half of which comes due in 2016.[20] The value of radio stations themselves has also dropped precipitously: in the heyday of consolidation following the Telecommunications Act of 1996, station valuations averaged about 15 times their yearly revenue potential. That multiple is now down into the single digits, with no real signs of recovery.[21] In 2008, for the first time since the passage of the Telecom Act, the annual aggregate value of radio station transactions dipped below $1 billion.[22] More than a decade since its founding, iBiquity receives the vast majority of operating revenue from venture capital firms and investment banks, and reported just $21 million in sales in 2011.[23]

All of these factors have put a strain on the ability of broadcasters to invest in HD Radio station conversions. *Radio World* reported in 2009 that many executives in charge of capital expenditures at broadcast conglomerates were delaying their digital conversion campaigns.[24] Capital cutbacks were so fierce that one engineer called it tantamount to "turning off a light switch" while another quipped, "If I was . . . preparing my budget, I'd put it all into duct tape."[25] Beasley Broadcast Group chief technology officer Mike Cooney confirmed that his company was backing down on "HD conversions in the small markets and . . . putting money more in things that have a quicker return on investment,"[26] while Cumulus Media, the second-largest U.S. radio conglomerate, officially amended its license agreement with iBiquity "to reduce the number of planned [HD] conversions, extend the build-out schedule, and increase the license fees to be paid for each converted station." Cumulus didn't disclose this change of heart on HD until 2012, in its annual report to the Securities and Exchange Commission. "At this juncture, we cannot predict how successful our implementation of HD Radio™ technology within our platform will be," said the company, "or how that implementation will affect our competitive position."[27]

The HD Radio Alliance has also curtailed its marketing efforts, committing just $110 million worth of radio advertisements in 2011—less than half of what broadcasters had pledged before the economic downturn. The Alliance has also rescinded many of the restrictions it imposed on member-stations, such as multicast format coordination, spot-load limitations on multicast channels, and even firm commitments of advertising inventory for promotional activities.[28]

Transmitter manufacturers report that inquiries about HD-compatible equipment have been few and far between following the market crash of 2008.[29] According to Tim Bealor, vice president of sales for Broadcast Electronics, "Unless we can figure out a way for broadcasters to make back their investment, [HD adoption] may be a futile effort."[30] Mike Troje, sales manager for Continental Electronics, echoed this sentiment. "It's a task to come up with what the right responses are for the industry when we don't know what the end game is."[31]

Public broadcasting's investment in HD Radio also took a significant hit when the National Telecommunications and Information Administration zeroed out the Public Telecommunications Facilities Program (PTFP) in 2011.[32] Many early-adopter noncommercial stations relied heavily on PTFP money to finance their HD installations. Coupled with the Corporation for Public Broadcasting's discontinuation of digital conversion grants, noncommercial broadcaster demand for new HD stations has effectively disappeared. Dennis Haarsager, NPR senior vice president of resources and technology, told *Current* in 2010, "HD Radio is neither DOA, nor is it assured of success. Rather, it has a plausible shot at market acceptance if we're patient and learn from our disruptors. Disruptive technologies don't need to overtake radio, they just have to skim our margin—and margins are pretty skinny where they exist at all."[33] But Brian Sanders, the engineer for 13 public radio stations in northern Arizona, has concluded that HD Radio is simply "not viable here . . . and [we] have serious doubts as to whether HD will ever be popular with listeners, even in more populated areas."[34]

In hopes of resuscitating broadcaster adoption, iBiquity slashed the one-time licensing fee for radio stations in 2010 from an average of $25,000 to $12,500 and agreed to accept payment in installments.[35] That September, iBiquity and Citadel Media announced an initiative whereby stations could convert to HD on the barter system, exchanging advertising inventory for capital expenditure, including the initial license fee.[36] Since Citadel is fronting the money and receiving the advertising inventory, this arrangement calls into question the feasibility of iBiquity's broadcaster-license revenue stream. Regardless, these tactics have not worked: "Every station would have adopted it by now if . . . it wasn't [for] all the required license fees," observed Finger Lakes Radio Group president and managing partner Alan Bishop. "Where's my return on investment? I see none."[37] Kelly Wallingford, the president of Wallingford Broadcasting Company and a former president of the Kentucky Broadcasters Association, told *Radio Ink* in 2013 that iBiquity's license structure doomed HD "from the start":

> I was part of a large group gathered at our State Convention in 2006 . . . [for] a seminar about HD. When iBiquity and the equipment reps got to the anticipated cost for HD, I could see they had lost the audience . . . I boldly rose to my feet and asked the panel if they realized they were asking broadcasters for an investment in an unproven technology that would rival or exceed the original cost of a whole radio station. I then asked the question that sent the entire room into a frenzied standing ovation and propelled my name into Kentucky broadcasting infamy: "Why so greedy?"

"Instead of exciting us, they priced us out. Instead of getting us on board, they turned an apathetic ear," explained Wallingford. "They alienated a large segment of the stations that could have been their greatest allies. Here's the cold hard truth: Radio doesn't even like HD. I have said it all along and will say it again: Until all of radio starts getting excited about HD, no one will be excited about it."[38]

In 2011, an "unscientific online survey" of readers by the Clear Channel-owned trade publication *Inside Radio* reported that 44% believed HD Radio would never reach the "tipping point" where it might actually replace analog broadcasting.[39] Mike Vanhooser, the president of Dallas-based Nova Electronics, believes that the industry is now stuck with an "albatross," and has encouraged independent broadcasters to "take the albatross from around their neck and make their *analog* signal an engineering masterpiece . . . and put some programming worth listening to on it" [emphasis in original].[40] Energy-Onix president Bernard Wise declared in 2010 that he would boycott future industry conventions out of shame "to be associated with the NAB in their promotion of this inferior system."[41] Cherry Creek Radio CEO Joe Schwartz worried in 2012 that digital broadcasting would become the purview of only those who could afford it. "It would cost my company in excess of $10 million to go to an HD platform," he told *Radio World.* "That is not going to happen," and he thinks there are many other independent broadcasters "who probably feel the same way."[42]

THE HUNT FOR HD'S KILLER APPLICATION

No broadcaster has yet recouped their investment, much less turned a profit, on their HD broadcasts, although several of the technology's features remain in play. Multicasting, long considered by broadcasters to be the technology's most valuable new functionality, is woefully underutilized. Less than two-thirds of all FM-HD stations run at least one additional stream of digital-only content, and many of these tend to be derivatives of a station's main format, or consist of formats no longer considered viable on analog radio.[43] For example, jazz and classical music has disappeared from the FM dial over the last decade, but now they represent "new" program-niches available on HD.[44] Many broadcast conglomerates, such as CBS and Clear Channel, import distant signals from stations in other markets to rebroadcast as FM-HD feeds, or simulcast their local AM stations as a "new" digital FM channel. "Not using new local channels for new local programming ideas feels like a lost opportunity and might invite further criticism that radio is not putting its spectrum to best use for local service," worried *Radio World.*[45] The economic downturn and collective disinterest led National Public Radio to shutter all of its multicast syndication services in 2012. Eric Nuzum, NPR's vice president of programming, said their "strategic decision" to offer "placeholder" content for FM-HD "didn't catch on very well."[46] Some stations have tried leasing their multicast channels to third parties, ranging from foreign-language broadcasters to condom brands and professional hockey teams.[47] No reliable list or database of available multicast channels exists, given the transitory commitment that most broadcasters make to them.[48]

Perhaps the state of multicasting is best illustrated on the *analog* FM dial, where broadcasters are using low-power FM translator stations in order to simulcast the content of their digital streams.[49] Cumulus Media has been the most aggressive at positioning these simulcasts not as enticements for listeners to try HD Radio itself, but rather as entirely new analog program outlets.[50] Many other broadcasters are now embracing this practice, creating a lucrative shadow-market for FM translator stations, which now regularly sell for tens or hundreds of thousands of dollars apiece. Furthermore, translators do not factor into the FCC's media ownership regulations, so this practice allows broadcasters to effectively circumvent local radio ownership caps. This is not the sort of innovation that seems to advance the cause of HD Radio in any meaningful way, though it does provide a clever method by which broadcast-adopters can recoup some of their investments in the technology.

In addition to multicasting, new datacasting features are also being promoted by HD proponents, who hope they will increase broadcaster interest and investment in HD Radio. This includes "Artist Experience" (AE), iBiquity's fancy branding moniker for radio with pictures. FM-HD stations equipped with AE functionality can broadcast still images and text to compatible digital receivers, including album art as well as images linked to advertisements. iBiquity CEO Struble thinks AE can be especially valuable to advertisers: "There's no reason when that Geico ad is playing that I shouldn't see the picture of the lizard on my radio screen."[51] However, AE implementation is apparently quite cumbersome: Paul Shulins, the chief engineer of Greater Media's Boston station cluster, told *Radio World* in 2012 that it took several months to get AE up and running on just one station—though subsequent implementations have gone much more quickly. Even so, enabling AE required a complete software rebuild of the iBiquity transmission chain and was not without its bugs. For example, AE-related imagery, which is served by a third party, was not always consistent or correct: "Either the images provided were too large, or not the most appropriate one . . . In some cases even profane or indecent images that were not appropriate for broadcast were returned. These obviously needed to be weeded out and handled on a case-by-case basis. Also getting the timing correct was a bit touchy." Coordinating these efforts between Greater Media, iBiquity, and the AE image-purveyor "proved to be time-consuming."[52] As of 2013, only "some 450 stations" (26% of all reported FM-HD adopters) have implemented Artist Experience, which requires broadcasters to sign an additional licensing agreement, and there's little information on the feature's residual costs.[53]

HD proponents also hope that song-tagging will catch on. Some FM-HD receivers allow a listener to bookmark a song they hear on the radio for later purchase via the iTunes Music Store. Apple pays broadcast partners approximately 5% of each purchase, which adds up to just pennies per transaction and will only be profitable if economies of scale are achieved. However, as of 2010, only nine broadcast conglomerates had committed to providing tagging functionality.[54] Further complicating matters, most HD Radio receivers don't have Internet connectivity, so the actual purchase of tagged songs must occur through another device, such as a computer or smartphone, which reduces the feature's convenience factor. Industry analysts such as Bishop Chen with Wells Fargo Securities think iTunes tagging "is way too new, and too arcane right now to be a driver of any investor sentiment" in HD Radio.[55]

The most "promising" of HD's datacast features actually has nothing to do with the primary mission of a radio station: the transmission of audio content. Many large broadcasters believe that the provision of traffic and weather data will be the technology's saving grace. The Broadcaster Traffic Consortium (BTC), an alliance of several major radio conglomerates, many of which are investors in iBiquity, is leading this effort. Overseen by Emmis Communications senior vice president and chief technology officer Paul Brenner (and based out of Emmis's corporate headquarters in Indianapolis), the BTC is "explicitly" an "HD Radio data distribution consortium business model, and one of its stated goals is to unify the industry for delivery of HD data services."[56]

In 2012, the BTC claimed membership by "20 commercial and noncommercial radio organizations representing about 1,500 stations, more than half of them digital, [with] close to 200 stations active in approximately 90 markets in the United States and Canada."[57] Traffic data payload on BTC-participant stations is pulled from the Internet and uses a minimum of 13 kilobits per second of digital bandwidth (or about 10% of the total data payload of an extended-hybrid FM-HD station), which can be doubled if the station desires more capacity.[58] Members of the consortium "share revenue according to a formula based on Arbitron population coverage; with a higher level of bandwidth commitment, a member's level of revenue rises."[59]

The BTC will not disclose the size of its revenue stream, and unity in this datacasting marketplace is disrupted by Clear Channel's own Total Traffic Network, which provides HD traffic data and weather information supplied by The Weather Channel on some 350 stations.[60] Proponents tout the fact that aftermarket traffic navigation devices are implementing HD reception capabilities, but these devices remain a marketplace minority.[61] Emmis's Brenner hopes that some auto manufacturers will eventually replace their satellite telematics systems with HD datacast reception capability, as such a shift would represent "a ginormous strategic win" for HD Radio.[62]

Still, it remains to be seen whether the popularity of any ancillary data service can drive demand for digital broadcasting itself. *Radio World* reported in 2012 that just 400 stations—or fewer than 20% of all HD broadcasters— offer any sort of advanced data services at all.[63] If broadcaster adoption of HD has been abysmal, and the use of its "extensibility" even more so, what does that say about the inherent vitality of the technology itself?

THE RECEIVER MARKETPLACE

Another critical element of radio's digital transition is the availability of HD receivers. Although consumer electronics manufacturers gradually backed away from direct opposition to the technology during the early stages of public policy development, they have resisted actually making receivers. In 2010, iBiquity reported that just 200,000 receivers had been sold between 2002 and 2008.[64] With an installed base of approximately 700 million radio receivers in the United States, this represents a penetration rate of just .03%. By the end of 2012, iBiquity claimed total sales had jumped to 12 million, or a penetration rate of 1.7%.[65]

The market for new standalone radio receivers, analog or digital, has been in decline since the turn of the century. Radio is no longer a medium for which you buy a special device; it's now just one of many content-delivery systems in a multi-platform media consumption routine. But even though the relative importance of this marketplace is waning, HD Radio's near-invisibility is notable. In 2009, iBiquity claimed that it had signed licensing agreements with some 140 companies to develop HD Radio receiver components and products,[66] but the gear has not appeared. Just a handful of table-top HD receivers exist, and there is only one portable model—produced by Insignia, the in-house electronics manufacturer for the Best Buy store chain.

In 2008, J. R. Russ toured electronics stores in Philadelphia and found a paucity of HD receiver choices; most of those he did manage to track down weren't even plugged in for demonstration purposes, and the stores' sales staff knew nothing about the technology. "How can broadcasters expect the public to get excited about HD Radio when it can't be uniformly heard everywhere?" asked Russ. "How can retailers sell the product if they don't stock it. Or if the products they do have on hand can't be demonstrated and employees are clueless? How can manufacturers justify building the product if it doesn't sell for these reasons?"[67] Aaron Read, who works at an FM-HD station in upstate New York, repeated the experiment in 2009. "Most stores don't carry any receivers at all," he reported. "The few that do, like RadioShack and Best Buy, often 'hide' them in a distant corner and don't hook up the antenna to the display unit . . . Given the overall lack of originality in [HD] formats, I fear there is little compelling reason for a listener to invest [in] an HD Radio in our area."[68] This lamentation remained in 2012: "Even online, the pickings are slim and carry a hefty price tag. Stores like Best Buy and Wal-Mart say they have HD Radios. Not in my neighborhood," contract engineer Mike Payne told *Radio World*. "I do work for a public radio network, and they run HD, but they are talking about turning it off, as there is only a handful of people using it."[69]

The growth in HD Radio's receiver-base over the last few years is primarily due to passive uptake in the automotive sector. In 2012, iBiquity's Struble claimed that an HD-equipped vehicle was sold every 15 seconds in the United States. However, considering that more than 14 million new vehicles were purchased in 2011, this means that only one in every seven came with HD onboard.[70] iBiquity also touts the fact that 33 automotive brands offer HD Radio,[71] but only six—Bentley, BMW, Mini, Rolls Royce, Scion, and Volvo—include it as standard equipment in all of their cars. The rest of the automotive industry offers HD receivers as part of "infotainment" options packages in select makes and models, if they offer it at all. For example, Toyota includes HD Radio as part of its Entune system package, which also provides the capability to stream Internet-delivered audio services.[72] Similarly, Ford's SYNC system includes HD as well as the ability to tether smartphones to the dashboard via Bluetooth.[73]

In fact, the dominant trend among automakers and the consumer electronics companies that support them involves adding Internet connectivity to cars and trucks at a rate much faster than adding HD functionality.[74] Pandora, a relative newcomer to the dashboard, boasts inclusion in more then 20 vehicle brands and has already surpassed HD in the aftermarket receiver space.[75] Jacobs Media and the Public Radio Program Directors Association reported in 2012 that just 6% of listeners had access to HD in their cars (a figure unchanged since 2010), while nearly 10% drove vehicles that had streaming Internet reception capabilities.[76] Just 2% of all radio listening in vehicles occurs via HD.[77] A 2013 survey of radio executives, group owners, and general managers found that 80% believe broadband in the car is the "single biggest threat" to the future of traditional broadcasting.[78] Simply put, in the new glass dashboards of modern automobiles, HD Radio already occupies an ancillary role.[79]

iBiquity's Struble first acknowledged in 2010 that radio's place of primacy in vehicles was at risk, and it would have to "fight for its position."[80] Just a year later, he told *Radio World*, "There was a monopoly position in the dash. It isn't [there] anymore and it's only getting worse."[81] In 2013, the publication declared the dashboard "a wild, wild west" of competing technologies.[82] iBiquity now predicts that HD won't become "standard on most models" until 2018 at the earliest.[83] Many broadcasters and industry observers agree that without meaningful uptake in automobiles, HD Radio's chances for survival are slim.[84] And unlike satellite radio services, which directly subsidized the placement of receivers in cars, iBiquity cannot afford to offer any incentives to auto manufacturers—though the broadcaster-driven HD Radio Alliance did give Ford some free advertising to get the technology into the SYNC system.[85]

Actual automotive marketing and support for HD Radio seems to be nonexistent. Broadcast engineer Tom Ray III, who oversaw the digital conversion of Buckley Broadcasting's WOR-AM in New York, was frustrated in 2010 when he bought a new Ford and assumed that HD Radio was a standard option:

> Lisa, the sweet sales person, promptly pointed to the Sirius sign and said that the car came with Sirius. Um, no, I told her. I said HD Radio. She got the head of the parts department. Nope—never heard of it. She got the head of the service department. Nope—never heard of it. She brought over the owner of the dealership, who went to his office and came back with what amounted to a ream of paper. It contained nothing about HD Radio from Ford. I drove off the lot the next day with the factory AM/FM radio that came with the car.[86]

iBiquity publicly acknowledged Ray's experience, but called it a "road bump" in Ford's "rolling launch" of the technology.[87] Further complicating matters, a legal firm is exploring a class-action lawsuit on behalf of luxury vehicle owners regarding the skittish reception and dubious "improved" audio quality of HD signals.[88] The lawyers suggest they may target vehicle manufacturers, not iBiquity, on the grounds of deceptive advertising.[89] Should this probe gain any traction, it could kill what tepid enthusiasm the auto industry has for HD Radio.

The other key receiver-space to watch for HD uptake is in mobile phones. Although this frontier in media consumption has been taking shape for the last decade or so, radio broadcasters have only recently twigged to its potential. However, many mobile phones can't receive radio broadcasts, and those models that have the hardware onboard often don't have the firmware or software necessary to utilize it. Instead of reaching out to phone manufacturers and wireless carriers to make the case for including such functionality, the National Association of Broadcasters first attempted to ram through a legislative mandate requiring it.[90] As a part of negotiations over the payment of performance rights fees in 2010,[91] the NAB tried to co-opt the music industry into the effort by tying proposed royalty rates to music industry support for a congressional "Radio Chip Mandate" covering both analog and HD Radio.[92]

Roundly excoriated by the music industry as an unnecessary complication to an already contentious issue,[93] the proposed mandate was also a non-starter for consumer electronics manufacturers. "As the representatives of an innovative and forward-looking industry, CEA will vigorously oppose any effort to force manufacturers by legislative fiat to include legacy technology in devices," wrote organization president Gary Shapiro in 2010. "Radio is a legacy horse and buggy industry trying to put limits on innovative new industries to preserve its format monopoly . . . We suggest you delete the technology mandates and recognize the free market works."[94]

Considering that the broadcast industry's political capital relative to telecommunications companies and consumer electronics manufacturers is much weaker now than it was just a decade ago, a government mandate for broadcast radio reception in mobile phones is highly unlikely, though it is an issue that broadcasters like to remind legislators of whenever possible.[95] This political misstep has forced broadcasters to carry the full burden of making the case for radio's adoption in mobile phones.

Much like HD Radio itself, broadcasters' overtures to the mobile phone market have overpromised and under delivered. At the 2010 NAB Radio Show, Radio Advertising Bureau president and CEO Jeff Haley told attendees, "HD Radio technology in a cell phone will be practical within about a year, creating additional services and revenue streams that become possible with a digital platform."[96] In 2011, iBiquity revised this outlook, claiming that "multiple handsets" from "several carriers" would be HD-ready by 2012.[97] Instead, 2012 proved to be a year of crash-development and furtive negotiation, as the NAB, Emmis Communications, and Intel forged a deal to design an HD Radio chip for mobile phones and develop an app to control it. The app, dubbed "NextRadio," serves up Artist Experience data and provides a back-end for interactivity via the phone's own data connection.[98] iBiquity and Emmis unveiled a prototype HD smartphone at the 2012 NAB convention, but it made few waves in the trades.[99] Emmis's Paul Brenner likened the accomplishment of building the smartphone prototype to "turn[ing] a dirt road into a gravel road"—a statement quite revealing about the embryonic state of broadcasters' engagement with mobile devices.[100]

While the cost of adding analog FM reception to a mobile phone is measured in pennies, the HD chipset currently costs between $2–$4 per unit, in large part due to the technology's proprietary nature.[101] Therefore, broadcasters are working to promote the adoption of analog FM in phones before pushing hard for HD functionality.[102] To that end, Emmis and Sprint signed a three-year agreement in late 2012 that commits Sprint to enabling FM reception in 30 million phones bundled with the NextRadio app.[103] But this comes at a hefty price for Emmis, which agreed to give $45 million in advertising inventory to Sprint, as well as 30% of the revenue derived from the sale of advertising delivered through NextRadio. In addition, Emmis is on the hook for maintaining NextRadio's back-end infrastructure, and broadcasters who wish to fully utilize the app will have to pay licensing and maintenance fees to Emmis.[104]

Emmis is actively soliciting other broadcasters to shoulder the costs of the Sprint deal. CEO Jeff Smulyan has suggested that every radio station in the country donate $30,000 in advertising inventory to pay for the agreement, and he told his peers at the 2012 NAB Radio Show that broadcasters should be prepared to give wireless carriers "millions in subsidizing" to gain access to the mobile phone platform.[105] But can every station afford to make such an investment? Emmis's efforts only provide a foothold on the nation's #3 wireless carrier, with an aggregate phone commitment that, in comparison to the accelerating growth of smartphone adoption overall, borders on the symbolic. It is also important to note that this campaign does nothing to help beleaguered AM broadcasters reach a broader potential listening audience. Just how many more subsidies will be necessary for radio to reach meaningful penetration in mobile phones?

Furthermore, what exactly does NextRadio bring to the table? Its developers suggest the app has more potential to transform the radio *advertising* experience than the *listening* experience, which may be of dubious value to listeners themselves.[106] Radio consultant Mark Ramsey thinks that while providing opportunities for interactive advertising "is certainly a good thing . . . that's not the same as giving consumers more of what they want from radio in the first place. Nor is it technologically novel . . . Where's the element designed to make consumers excited enough to demand this and to talk it up with their friends? Indeed, is this made with consumers or broadcasters in mind?"[107] The bottom line is that HD Radio's place in mobile phones is far from assured—or, as is shaping up to be the case in automobiles, the commitment to the platform may turn out to be too little, too late.

LISTENERS TUNE OUT HD RADIO

Listener uptake of HD Radio also reflects a crippled state of affairs. Overall, radio listening fell 16% in the last decade, and 23% among listeners between the ages of 18 and 24.[108] In 2013, Arbitron reported that radio broadcasting still reached an impressive 242.8 million Americans weekly—but of those, just 5.4 million (2.2%) listen to HD signals.[109] Bridge Ratings published two studies in 2009 and 2010 that show radio listening in decline, while the use of other digital audio distribution systems experienced significant growth.[110] According to Bridge, terrestrial radio listening first reached a downward tipping point in 2002–2003, while the "collective momentum of digital [audio] alternatives" began to outpace radio in 2007. This suggests that HD Radio has always been an also-ran in the convergent listening environment.[111]

There is little fertile ground for new HD listenership. In a presentation at the 2008 NAB Radio Show, Jacobs Media general manager Paul Jacobs presented ethnographic research on the audio consumption habits of young adults. "Mild groans went up . . . when one respondent, asked whether she had a radio, said that she might have one 'in the top of my closet,' though it might not work."[112] Jacobs told broadcasters that he wanted to show segments demonstrating listener engagement with HD Radio, "but only two of them even knew what it was, and it was clear that they didn't know what they were talking about."[113] Overall, Jacobs reported that listener familiarity with HD Radio "dropped from 70 percent in 2007 to 60-some percent in 2008."[114]

Jacobs Media's audience results are not unique. In a 2009 survey of 30,000 public radio listeners, nearly half expressed awareness of HD Radio, but only 3% had actually bought an HD receiver. Of the rest, 11% said they were "very likely" to purchase one in the future, while 65% reported they were "somewhat" or "not at all likely" to.[115] Around the same time, Arbitron and Edison Research found that 66% of listeners surveyed were not interested in HD Radio, while just 6% were "very interested."[116]

In 2012—following several years of on-air promotion—the majority of American radio listeners still didn't know exactly what HD Radio really was. Just 54% of respondents to a Kassof survey had "heard of" HD Radio—a decline of 13% since 2008. Among those "in the know," 16% couldn't describe it, 20% said the main benefit of HD was better audio quality (though they made this inference primarily based on the technology's "HD" branding), and only 8% of respondents knew about FM-HD's multicasting feature. Six percent actually confused HD with digital satellite broadcasting.[117]

The decline of listener interest and engagement with HD Radio is likely to continue, especially as the popularity of other digital audio delivery services grows. In 2012, Jacobs Media conducted "the largest radio tech survey ever" involving more than 57,000 radio listeners, and found that four out of every ten also "listen to streaming radio on a weekly basis or more often." Among Pandora users alone, 43% consider the service to be "radio," while 49% do not.[118] Broadcasters might be able to take advantage of listeners' migration to other platforms by providing Internet streams of their programming, but in 2012 Arbitron reported that just 59% of radio stations did so.[119]

Many broadcasters, such as Clear Channel, are aggressively promoting streaming—so much so that the radio stations Clear Channel owns have been effectively repositioned as promotional outlets for its iHeartRadio online platform.[120] Several other commercial and noncommercial broadcasters, such as Cox, Cumulus, the Educational Media Foundation, Emmis, Greater Media, Univision, and a variety of public and college radio stations, have also agreed to offer their stations exclusively through iHeartRadio, making Clear Channel a powerful online aggregator of broadcast-streaming.[121] But others are backing away from the Internet: Saga Communications announced in 2012 that it would stop streaming its radio stations located outside the top 100 U.S. radio markets, and restrict online listening to people who live within the stations' actual coverage areas. Saga also implemented a 90-minute "listening window" on radio streams, requiring users to click through if they wish to listen longer. CEO Ed Christian told the trades that the company spent about $800,000 per month providing online streams before the cutback—a cost that Christian suggested wasn't worth the return on investment.[122]

Media buyers are increasingly concerned that the radio industry's general ignorance of the Internet is turning both listeners and advertisers away. Brad Bernard, the vice president of online media and analytics for Harmelin Media, thinks that if broadcasters are "going to have credibility in the online space, they need to learn the lingo better . . . [they're] using the [digital] language of five or six years ago," referring to things like "hyperlinks" which most marketers (and Internet users, for that matter) now take for granted.[123] If broadcasters lack a fundamental engagement with how the Internet is changing their industry, how can they reasonably expect to entice listeners to stay with traditional "radio," much less HD, especially if the migration to other platforms that provide radio-like services is already well underway? Like newspapers and television before it, radio's navigation of the convergence phenomenon has sparked an extended period of confusion and anxiety among broadcasters as they come to grips with the new shape of our modern media environment. By all accounts, the industry's continued investment in HD Radio only seems to complicate this process.

It is fair to say that HD Radio is on life support in the United States. Even after getting the FCC to authorize additional compromises to analog signal integrity in order to improve HD's basic functionalities, most broadcasters, receiver manufacturers, and listeners remain unwilling to invest in the technology. The majority of U.S. broadcasters have chosen to opt out of HD Radio altogether, while those that opted in are struggling mightily to find any metric by which to show that the technology has some viability. The prevailing strategy of proponents is to push forward with several new "advancements" in HD technology in hopes that one of them will stick. Relative to its newfound competition, HD Radio seems lackluster to all of the core constituencies it needs to survive. Regulators have decreed that marketplace forces would guide radio's digital transition—but if this is so, the marketplace does not bode well for HD Radio.

Ultimately, it will be the U.S. marketplace that decides the technology's fate, as HD Radio has no meaningful footprint internationally. The only other countries to formally adopt HD Radio are Jamaica, Mexico, Panama, Puerto Rico, and the Philippines—nations well within the U.S. sphere of trade influence, and none of which have the economic heft to influence radio markets elsewhere.[124] Canada allows limited HD operations on stations along the U.S. border, primarily to allow for datacasting service continuity, but shows little interest in full-scale adoption.[125] Brazil has similarly flirted with HD Radio, but doesn't plan to make any decision until a comparative analysis of all possible digital broadcast technologies has been completed. Generally speaking, Latin American countries have refused to commit to any digital transition for radio until they see how the technologies fare in their home markets.[126] iBiquity Digital Corporation claims that it has launched "joint venture operations" and "limited operation" in several countries, and sees "advanced interest" in many more, but none have moved beyond the experimental stage, and many "tests" have been novelties at best, especially in Europe and Asia.[127] Incidentally, iBiquity does not charge broadcaster-licensing fees outside the United States, so international disinterest ultimately stems from the technology's other fundamental deficiencies.[128]

The malaise of radio's digital transition has provoked a profound level of introspection among broadcasters about their own self-worth. "Will [radio] fight fire with fire and move to selected microformats, or come full circle and return to a variety schedule?" Skip Pizzi wondered in 2009. "Or could some form of hyperlocalism find its way back into broadcasters' DNA, emerging from an almost forgotten chromosome? . . . At the moment, everyone acknowledges these questions, but no one has many answers."[129] Consensus is developing that suggests the bridge to a sustainable position for radio in a convergent media environment will be through devices and platforms that do not involve its traditional broadcast infrastructure.[130] A 2011 survey of radio professionals by studio equipment-maker Wheatstone found that 61.3% of respondents from standalone stations and 54.8% of respondents from group-owned stations expect the majority of their listeners to come from online—not over the air—within 15 years.[131] Skip Pizzi advised that HD Radio be seen as "the long-term, speculative play in the portfolio, and balance it with other components on a faster and more likely track to new media [return on investment]."[132] Shortly after so eloquently expressing the radio industry's doubts regarding its digital future, the NAB hired Pizzi as its "director of digital strategies."[133] However, since he took the job in 2010, Pizzi has not made any further public statements about HD Radio and its prospects.

Many others within the industry are much less hopeful. "Everyone can read the tea leaves at this point about HD Radio," consultant Mark Ramsey explained to *Current* in 2012. "You can pretty much pick your cliché about it—the goose is cooked, the horse has left the barn, or the ship has left the harbor." The multibillion-dollar effort spent on HD development and promotion has "been an astonishing waste of previous industry resources and audience attention" that diverted time and money away from investments in quality programming—the one thing that has been proven to generate listener interest. Ramsey now believes that HD Radio was "fundamentally ill-conceived from day one, and now they're trying to reverse-engineer it to find something that fits the technology."[134] But Amy Mitchell, director of the Pew Project for Excellence in Journalism, thinks that "technology has really moved beyond" what HD Radio itself has to offer.[135]

What will the future landscape of "radio" look like, and where will traditional broadcasters fit into the digital milieu? These questions represent the core of radio's digital dilemma, and they transcend bounded considerations of any particular technology, business model, or audience metric. The longer that incumbent broadcasters neglect or avoid engagement with this dilemma, the more likely it becomes that "radio" as we've known it will be subsumed by the convergence phenomenon, leading to the potential loss of values and attributes that made the medium unique and useful, and with greatly reduced agency to shape its own destiny.

NOTES

1. See Paul McLane, "Radio, Everywhere and on Everything," *Radio World*, June 6, 2012, 4, 14.
2. Federal Communications Commission, "Broadcast Station Totals as of June 30, 2013," *FCC.gov*, July 10, 2013, http://www.fcc.gov/document/broadcast-station-totals-june-30-2013.
3. Barry McLarnon, "AM IBOC Stations On The Air," *Topaz Designs*, last modified June 7, 2013, http://topazdesigns.com/iboc/station-list.html.
4. Leslie Stimson, "Radio Eyes All Forms of 'Digital,'" *Radio World*, October 24, 2012, 10.
5. Guy Wire, "Take a Bigger Slice of Net Income Pie," *Radio World Engineering Extra*, June 11, 2006, 35.
6. See Tom Ostenkowsky, "Sunday in One Word: Digital!" *Radio World*, March 24, 2010, 22, 24–25; Leslie Stimson, "For Digital Radio, 'Hardware Sells Hardware'," *Radio World*, June 2, 2010, 1, 3; and Michael LeClair, "NRSC Revises HD Radio Standards," *Radio World*, December 12, 2012, 3.
7. Ostenkowsky.

8. Randy J. Stine, "Among AM HD Users, Opinions Vary," *Radio World*, September 2, 2010, http://www.rwonline.com/article/among-am-hd-users-opinions-vary/3777; see also Leslie Stimson, "AM HD Radio Has Stalled. Now What?" *Radio World*, August 31, 2010, http://www.rwonline.com/article/am-hd-radio-has-stalled-now-what/3774.

9. Paul McLane, "Downs Advocates for AM Solutions," *Radio World*, November 16, 2011, 4, 16.

10. Leslie Stimson, "All-Digital AM Tests Considered," *Radio World*, August 15, 2012, 3, 5.

11. Leslie Stimson, "Station Chosen for All-Digital Test," *Radio World*, October 10, 2012, 8.

12. Ibid., 5.

13. John Anderson, "All-Digital AM-HD Tests Underway," *DIYmedia.net*, December 19, 2012, http://diymedia.net/2012/12/19/all-digital-am-hd-tests-underway/.

14. Leslie Stimson, "More All-Digital AM Tests Planned," *Radio World*, March 12, 2013, http://radioworld.com/TabId/64/Default.aspx?ArticleId=218248.

15. Leslie Stimson, "Initial All-Digital AM IBOC Tests Completed; Results Being Digested," *Radio World*, January 17, 2013, http://radioworld.com/article/initial-all-digital-am-iboc-tests-completed-results-being-digested/217304.

16. "Pai: Time to Review All AM Rules," *Radio World*, October 24, 2012, 5, 8.

17. Leslie Stimson, "Station Chosen for All-Digital Test," *Radio World*, October 10, 2012, 6.

18. James E. O'Neal, "IBOC at Night, Five Years Later," *Radio World*, February 13, 2013, 36–37.

19. Quoted in Karen Everhart, "AM Proves to be a Hard Sell, Even for News Radio," *Current*, August 9, 2010, http://www.current.org/wp-content/themes/current/archive-site/tech/tech1014-expansion-and-AM.shtml.

20. See Chris Nolter, "Clear Channel to Attack 2014 Debt First," *The Deal Pipeline*, February 28, 2013, http://www.thedeal.com/content/tmt/clear-channel-to-attack-2014-debt-first.php; and "Radio Ad Projections," *Radio World*, March 11, 2009, 24.

21. "Radio Ad Projections," 29.

22. "Transactions Seek Lows," *Radio World*, May 6, 2009, 25.

23. Ben Mook, "Slow Growth for HD Radio," *Current*, November 5, 2012, http://www.current.org/2012/11/slow-growth-for-hd-radio/.

24. Randy J. Stine, "Surprise: Cap-Ex Will Be Taut in 2009," *Radio World*, January 1, 2009, 1, 8.

25. Paul McLane, "Radio, Back Where It Belongs," *Radio World*, May 5, 2010, 4.

26. Leslie Stimson, "For Cooney, It's About Return on the Dollar," *Radio World*, March 25, 2009, 6.

27. Cumulus Media, Inc., *Form 10-K For the Fiscal Year Ended December 31, 2011*, March 12, 2012, via United States Securities and Exchange Commission, http://www.sec.gov/Archives/edgar/data/1058623/000119312512110079/d309291d10k.htm, 23–24.

28. "News Roundup," *Radio World*, December 15, 2010, 10.

29. Paul McLane, "From the Editor: 'The Quantity Was Down, But . . . ,'" *Radio World*, May 20, 2009, 4.

30. Quoted in Leslie Stimson, "FCC Ready to Move on Power Increase?" *Radio World*, October 21, 2009, 14.

31. Quoted in ibid.

32. Leslie Stimson, "PTFP Announces Its Own Demise," *Radio World*, June 15, 2011, 6.

33. Dennis Haarsager and Mike Starling, "What Public Stations Should Consider About Upgrading HD Power," *Current*, May 17, 2010, http://www.current. org/wp-content/themes/current/archive-site/tech/tech1009hdradio.shtml.
34. Randy J. Stine, "Sanders Keeps KNAU On the Air," *Radio World*, May 19, 2010, 6.
35. "News Roundup: Licensing," *Radio World*, March 1, 2010, 7.
36. See "iBiquity, Citadel Media to Facilitate Digital Projects," *Radio World*, September 27, 2010, http://www.rwonline.com/article/ibiquity-citadel-media-to-facilitate-digital-projects/5237; Joseph Palenchar, "HD Radio Outlines New Initiatives," *This Week in Consumer Electronics*, September 28, 2010, http:// www.twice.com/news/hd-radio-outlines-new-initiatives-0; and Leslie Stimson, "HD Radio on the Side," *Radio World*, November 3, 2010, 10.
37. Quoted in John Merli, "Radio, the 'Always With You' Medium," *Radio World*, September 8, 2010, 34.
38. Kelly Wallingford, "The Future of HD Is in the Hands of Independents," *Radio Ink*, March 25, 2013, http://www.radioink.com/Article.asp?id=2632826& spid=24698.
39. "Poll: HD Radio Tipping Point is Years Away," *Inside Radio*, January 2011, http://www.insideradio.com//Article.asp?id=2072825&spid=32061.
40. Mike Vanhooser, "I Told You So," *Radio World*, October 21, 2009, 34.
41. Bernard Wise, "A Vote of No Support," *Radio World*, March 24, 2010, 69.
42. Quoted in Leslie Stimson, "Radio Eyes All Forms of 'Digital,' " 12.
43. iBiquity reported in 2013 that 1,146 stations were multicasting with at least two program streams (the station's primary program content plus a digital-only service); see Leslie Stimson, "iBiquity: Pew Data Isn't Accurate," *Radio World*, March 19, 2013, http://www.radioworld.com/article/ ibiquity-pew-data-isn%E2%80%99t-accurate/218396.
44. Mook.
45. "Is It the Best We Can Do? CBS Rebroadcast Strategy Invites Scrutiny," *Radio World*, December 16, 2009, 34.
46. Quoted in Mook.
47. See Leslie Stimson, "Making Money with HD2, HD3," *Radio World*, April 14, 2010, http://web.archive.org/web/20110910020231/http://www. nabshowdaily.com/NabShowToday/99120; and Michael LeClair, "Time to Look Our Best," *Radio World Engineering Extra*, February 20, 2013, 4.
48. It was big news when an FM station in St. Louis dedicated full-time talent to program its HD multicasts in 2013; see Joe Holleman, "Radio Rich Off Air to Concentrate on Emmis HD Channels," *St. Louis Post-Dispatch*, January 30, 2013, http://www.stltoday.com/lifestyles/columns/joe-holleman/ radio-rich-off-air-to-concentrate-on-emmis-hd-channels/article_a9b3619b-895e-5e39-aafd-ed13e12997be.html.
49. An FM translator station is a secondary broadcast service. Translators are limited to an operating power of 250 watts or less, may not originate their own programming, and must accept interference from full-power radio stations. See Federal Communications Commission, "FM Translators and Boosters," *FCC Encyclopedia*, September 13, 2011, http://www.fcc.gov/ encyclopedia/fm-translators-and-boosters-general-information.
50. See "Translator/HD2 Simulcast Keeps 'Touch' on FM in Harrisburg," *All Access*, August 27, 2008, http://www.allaccess.com/net-news/archive/ story/45037/translator-hd2-simulcast-keeps-touch-on-fm-in-harr; Kara Scheerhorn, "WMHW HD2 Station Launching on 101.1 FM," *Central Michigan Life*, November 20, 2009, http://www.cm-life.com/ 2009/11/20/wmhw-hd2-station-launching-on-101-3-fm/; and Lance Venta,

"103.7 The Dam Debuts in Kansas City," *Radio Insight*, March 18, 2010, http://radioinsight.com/blog/headlines/netgnomes/3196/cumulus-launches-new-kansas-city-fm/.
51. Quoted in Leslie Stimson, "'Battle For the Dash' Is On," *Radio World*, March 1, 2011, 6.
52. Paul Shulins, "An Early Experience With Artist Experience," *Radio World*, November 21, 2012, 1, 6, 8.
53. See "Most Car Brands Offer HD Radio," *Radio World*, April 4, 2013, http://www.radioworld.com/article/most-car-brands-offer-hd-radio/218779, and Stimson, "HD Radio on the Side," 8.
54. Randy J. Stine, "What's the Outlook for Tagging?," *Radio World,* May 5, 2010, 5.
55. Ibid., 6.
56. Paul McLane, "He Emphasizes 'Innovation' Over 'Legacy,'" *Radio World*, November 21, 2012, 4.
57. See ibid., and Leslie Stimson, "The Leslie Report," *Radio World*, February 1, 2013, 3.
58. The AM-HD system does not have the bandwidth to accommodate BTC datacasts. See Broadcaster Traffic Consortium, "How it Works," *Radio BTC*, 2010, http://www.radiobtc.com/how-it-works.html; and Leslie Stimson, "BTC Senses Renewed Interest in Radio," *Radio World*, August 1, 2010, 8.
59. Leslie Stimson, "BTC Senses Renewed Interest in Radio," 6.
60. Leslie Stimson, "iBiquity Sees CE Gains in 2013," *Radio World*, February 13, 2013, 10.
61. Leslie Stimson, "Radio/Audio Trends for the Holidays," *Radio World*, December 1, 2011, 5.
62. Paul McLane, "He Emphasizes 'Innovation' Over 'Legacy'," 5.
63. Leslie Stimson, "Radio Eyes All Forms of 'Digital'," 10.
64. Leslie Stimson, "For Digital Radio, 'Hardware Sells Hardware'," 8.
65. "News Roundup," *Radio World*, December 19, 2012, 10. 4.5 million of those were sold in 2012 alone; see Michael LeClair, "Time to Look Our Best," 4.
66. Leslie Stimson, "More Factories Licensed to Make Devices," *Radio World*, June 3, 2009, 6.
67. J. R. Russ, "Exploring HD Radio Availability in Philly," *Radio World*, November 19, 2008, 18, 20.
68. Aaron Read, "A Listening Test in the Finger Lakes," *Radio World*, July 1, 2009, 12.
69. Mike Payne, "Oh, HD Where Art Thou?," *Radio World*, November 21, 2012, 38.
70. Bernie Woodall and Ben Klayman, "Auto Industry Posts Best U.S. Sales Year Since 2007," *Reuters*, January 3, 2013, http://www.reuters.com/article/2013/01/03/us-usa-autosales-idUSBRE9020BN20130103.
71. "Thirty-Three Automakers Now Offer HD Radio," *Radio World*, April 3, 2013, http://www.radioworld.com/article/thirty-three-automakers-now-offer-hd-radio/218758.
72. Darren Murph, "Toyota Entune In-car Infotainment System Interfaces With Your Smartphone, Does Everything but Drive," *Engadget*, January 4, 2011, http://www.engadget.com/2011/01/04/toyota-entune-in-car-infotainment-system-interfaces-with-your-sm/.
73. See Wayne Cunningham, "Toyota Integrating Internet Radio App," *CNet*, November 15, 2010, http://reviews.cnet.com/8301-13746_7-20022839-48.html; and Ford Motor Company, "Ford SYNC," 2012, http://www.ford.com/technology/sync/.

74. Karl Henkel, "Automakers Tuning Out Traditional In-car Radios," *The Detroit News*, April 29, 2013, A1.
75. Leslie Stimson, "CE Manufacturers Embrace Connectivity," *Radio World*, February 13, 2013, 5, 8.
76. Mook.
77. Ibid.
78. Mark Kassof, "Car Wars: The Big Challenge," *Kassof & Co.*, March 6, 2013, http://kassof.com/2013/car-wars-the-big-challenge/.
79. See Mark Ramsey, "What 4G LTE Internet Access in Your Car Means for Radio," *Mark Ramsey Media*, February 25, 2013, http://www.markramsey media.com/2013/02/what-4g-lte-internet-access-in-your-car-means-for-radio/.
80. Bob Struble, "The Race for the Dashboard Is On," *Radio World*, January 1, 2010, 18.
81. Leslie Stimson, " 'Battle For the Dash' Is On," *Radio World*, March 1, 2011, 5.
82. Leslie Stimson, "Dashboard Has Become a 'Wild West,' " *Radio World*, May 22, 2013, http://www.radioworld.com/article/dashboard-has-become-a-%E2%80%98wild-west/219531.
83. Leslie Stimson, "HD Radio on the Side," 26.
84. See Michael LeClair, "Time to Look Our Best," 4; and Mook.
85. Leslie Stimson, "HD Radio Awareness Remains Low," *Radio World*, February 15, 2012, 3.
86. Thomas R. Ray III, "HD Radio Shouldn't Be This Hard," *Radio World*, August 11, 2010, http://www.rwonline.com/article/hd-radio-shouldn39t-be-this-hard/3684.
87. Robert Struble, "Ray Experienced a Road Bump," *Radio World*, September 8, 2010, 3, 5.
88. Leslie Stimson, "The Leslie Report," *Radio World*, September 22, 2010, 7.
89. See Keefe Bartels, "HD Radio Not High Definition: Investigation by Keefe Bartels," 2012, http://www.keefebartels.com/content/hdradionothighdefinition investigationbykeefebartels.
90. Matthew Lasar, "Fear the Mobile Device Mandate Monster, Now With HD Radio Too!" *Radio Survivor*, October 27, 2010, http://www.radiosurvivor.com/2010/10/27/fear-the-mobile-device-mandate-monster-now-with-hd-radio-too/.
91. At present, broadcasters are exempt from paying performance royalties on the music they air. This historical exception is under attack as streaming media services gain popularity and radio's relevance to the success or failure of new artists wanes.
92. See NAB Radio Board, *Term Sheet for Performance Rights Agreement*, October 25, 2010, http://www.nab.org/documents/newsroom/pdfs/102510_NAB TermSheet.pdf.
93. See Eli Center, "NAB Rewrites July Agreement and Undermines Economics of Compromise," *musicFIRST Coalition*, October 26, 2010, http://www.music-firstcoalition.org/?page=blog_index&postid=1381399.
94. Gary Shapiro, CEA Letter to NAB, October 26, 2010, http://static.arstech nica.com/NAB%20BoardMeetingLetter.pdf.
95. See United States Congress, House of Representatives, Committee on Energy and Commerce, Subcommittee on Communications and Technology, "The Future of Audio," June 6, 2012, http://energycommerce.house.gov/hearing/future-audio; and Anderson, "Beware Broadcasters' Post-Sandy Opportunism," *DIYmedia.net*, November 29, 2012, http://diymedia.net/2012/11/29/beware-broadcasters-post-sandy-opportunism/.
96. Leslie Stimson, "They'll Address the Floor," *Radio World*, November 3, 2010, 5.

97. See Leslie Stimson, "Radio Talks EAS, LPFM, Translators," *Radio World*, October 19, 2011, 10; and "News Roundup," *Radio World*, November 16, 2011, 8.
98. See Leslie Stimson, "Intel Acquires SiPort," *Radio World*, July 13, 2011, 3; Leslie Stimson, " 'Rapid Deployment' Challenges Radio," *Radio World*, May 23, 2012, 6; and "News Roundup," *Radio World*, November 21, 2012, 19.
99. Paul Riismandel, "HD Radio Smartphone Prototype Lands with a Thud," *Radio Survivor*, April 26, 2012, http://radiosurvivor.com/2012/04/26/hd-radio-smartphone-prototype-lands-with-a-thud/.
100. Leslie Stimson, "Sprint Deal Is a Building Block," *Radio World*, February 13, 2013, 1, 3.
101. See Leslie Stimson, "Radio Eyes All Forms of 'Digital,' " 10; and Leslie Stimson, "Stations Work to Implement AE," *Radio World*, June 1, 2011, 14.
102. Stimson, "Sprint Deal is a Building Block."
103. NextRadio can also be configured to work with analog FM stations that use the Radio Data System (RDS).
104. See Leslie Stimson, "Sprint Deal Is a Building Block," and "Radio's Sprint Bill: Three-year Term," *Inside Radio*. February 2013, http://www.insideradio.com/article.asp?id=2616208.
105. See Leslie Stimson, "Emmis, Industry Preps for FM in Cellphones," *Radio World*, February 27, 2013, http://www.rwonline.com/article/emmis-industry-preps-for-fm-in-cellphones/217990; and Stimson, "Radio Eyes All Forms of 'Digital,' " 10.
106. See Stimson, " 'Rapid Deployment' Challenges Radio," 6; and Emmis Radio, *NextRadio*, 2013, http://www.tagstation.com/nextradio/.
107. Mark Ramsey, "As Long as FM Radio is on Sprint Phones . . .," *Mark Ramsey Media*, January 14, 2013, http://www.markramseymedia.com/2013/01/as-long-as-fm-radio-is-on-sprint-phones/.
108. Jeremy Egner, "Broadcasters, Meet Your Future," *Current* XXVII, no. 4 (March 3, 2008), 12.
109. Arbitron, "Radio Adds More Than 1.6 Million Weekly Listeners, According to the RADAR March 2013 Report," *Arbitron.com*, March 11, 2013, http://arbitron.mediaroom.com/index.php?s=43&item=866.
110. Bridge Ratings, "2009 Competitive Media Usage Overview Update," December 7, 2009, http://www.bridgeratings.com/press.12.07.09.CompMediaUse.html.
111. Bridge Ratings, "Terrestrial Radio's Run Through the New Media Gauntlet 1998–2010," October 20, 2010, http://www.bridgeratings.com/press.10.14.10.Gaunlet.htm.
112. Egner, 12.
113. Quoted in ibid.
114. See Leslie Stimson, "Digital Integrity Is Focus of PREC," *Radio World*, June 18, 2008, 6; and Leslie Stimson, "Public Radio Listeners Talk Tech," *Radio World*, May 6, 2009, 5.
115. Karen Everhart, " 'You're Losing Your Biggest Fans,' " *Current*, March 2, 2009, http://www.current.org/wp-content/themes/current/archive-site/web/web0904techpoll.shtml.
116. Editorial, "Consumer Awareness Is a Slippery Thing," *Radio World*, May 21, 2008, 46.
117. Leslie Stimson, "HD Radio Awareness Remains Low," *Radio World*, February 15, 2012, 3.
118. See Jacobs Media, "Jacobs Media's All-Format Techsurvey8 Released," *Jacobs Ideas & Articles*, June 2012, http://www.jacobsmedia.com/articles/ts8_release042712.asp; and Fred Jacobs, "Radio Dithers, Tim Smiles," *jacoBLOG*, June 20, 2012, http://jacobsmediablog.com/2012/06/20/radio-dithers-tim-smiles/.

119. Arbitron, *National Radio Format Shares and Station Counts, Fall 2011*, December 4, 2012, http://www.arbitron.com/downloads/Radio_Today_2012_execsum.pdf.

120. John Anderson, "When the Internet Takes Over Radio Stations," *DIYmedia. net*, January 31, 2013, http://diymedia.net/2013/01/31/when-the-internet-takes-over-radio-stations/.

121. See Ken Dardis, "It Hurts to Believe Your Own Hype," *Audio Graphics*, February 7, 2012, http://www.audiographics.com/agd/020712-1.htm; Michael Schmitt, "iHeartRadio's Reported Exclusivity Agreement Worries Some," *Radio and Internet Newsletter*, February 8, 2012, http://kurthan son.com/category/issue-title/rain-28-iheartradio%27s-reported-exclusivity-requirement-worries-some; Michael Schmitt, "B'dcasters Discuss Whether the Benefit of an iHeartRadio Listing Outweighs the Cost," *Radio and Internet Newsletter*, February 16, 2012, http://kurthanson.com/category/issue-title/rain-216-b%27dcasters-discuss-whether-benefit-iheartradio-listing-outweighs-cost; and Stimson, "CE Manufacturers Embrace Connectivity," 1, 5, 6, 8.

122. Michael Schmitt, "Saga to Limit Streaming to Top 100 Markets, Points to High Costs," *Radio and Internet Newsletter*, June 15, 2012, http://kurthan son.com/news/saga-limit-streaming-top-100-markets-points-high-costs.

123. Quoted in John Anderson, "Radio Advertisers' Digital Dilemma," *DIYmedia. net*, September 27, 2012, http://diymedia.net/2012/09/27/radio-advertisers-digital-dilemma/.

124. See Carlos Eduardo Behrensdorf, "Brazil Could Pick Digital Standard in 2010," *Radio World*, May 5, 2010, 10, 12; Jorge J. Basilago, "Digital Radio's Future Murky in Latin America," *Radio World*, March 9, 2011, 8; "News Roundup," *Radio World*, March 23, 2011, 8; and "News Roundup," *Radio World*, September 26, 2012, 3.

125. Leslie Stimson, "The Leslie Report," *Radio World*, July 14, 2010, 3, 8.

126. Basilago, 1, 8, 10; and "Brazil Digital," *Radio World*, February 1, 2013, 5.

127. See iBiquity Digital Corporation, "HD Radio™ Broadcasting Around the World," 2012, http://www.ibiquity.com/international/hd_radio_adoption_around_the_world, and "HD Radio Broadcasting Around the World," 2013, http://www.ibiquity.com/international.

128. Basilago, 10.

129. Skip Pizzi, "Ponder This Post-Broadcast Paradigm," *Radio World*, June 3, 2009, 24.

130. See Skip Pizzi, "Is the Perfect Storm Approaching?", *Radio World*, January 14, 2009, 18; and James Careless, "Radio Submits Its Applications," *Radio World*, July 15, 2009, 20, 22.

131. Josh Gordon, "What's Next in Radio Technology?" *Radio World*, February 9, 2011, 37–38.

132. Skip Pizzi, "Radio Applies the Pogo Principle," *Radio World*, September 23, 2009, 18.

133. "Skip Pizzi Joins NAB Science & Technology," *Radio World*, December 13, 2010, http://www.rwonline.com/article/skip-pizzi-joins-nab-science-amp-technology/4348.

134. Mook.

135. Quoted in ibid.

8 Confronting Radio's Digital Dilemma

Radio's digital dilemma is quite real, and the circumstances that have engendered it are now clear. What remains to be seen is whether the trajectory of radio's digital transition in the United States is amenable to proactive modification, and whether broadcasters themselves have the wherewithal to adapt to a convergent media environment irrespective of the technologies they ultimately use to get there. Initial attempts via the HD Radio system have not been successful, thanks to the inherent deficiencies of the technology itself and the lack of regulatory engagement with the real-world consequences of its design and implementation. Simply becoming "bit radiators" also does not address shifting expectations among the listening public about what "radio" in a digital environment actually is. In many respects, radio's digital transition represents a critical juncture of sorts for the medium itself: so long as broadcasters and regulators continue to address the phenomenon of convergence in such a single-dimensional fashion, this dilemma will only become more complicated and challenging.

THE PERILS OF DIGITAL RADIO POLICYMAKING

When ideology trumps science, seemingly rational actors can make arguments and decisions with a degree of ignorance and hubris that is painful to behold. Fundamentally, the broadcast industry's narcissistic oblivion regarding convergence itself, coupled with an FCC that was enamored with the phenomenon but had no clear understanding of what it meant, led to the deployment of a digital radio technology unsuited for the task at hand. Considering the amount of investment made in HD Radio to date, the technology is not likely to be abandoned voluntarily. One of the consequences of staying the course is the widening of legitimate debate regarding the future of terrestrial broadcasting from perspectives beyond those historically defined by the medium's primary actors. This could have constructive implications—but only if the dominant paradigm of media policymaking can be shifted away from a marketplace orbit. In a nutshell, the development of HD Radio is not a story of evil people doing stupid things, or stupid people doing evil things; such a simplistic characterization fails to indict the system of policymaking that allowed U.S. broadcasters to facilitate their own marginalization through the implementation of a digital broadcast technology fraught with problems.

The regulatory dysfunction of radio's digital transition was precipitated by broadcast incumbents who attempted to address the convergence phenomenon under conditions they could wholly control—in hindsight, a fool's errand. They prioritized increasing private control over the airwaves above all other possible outcomes. This has allowed both HD Radio proponents and regulators to ignore or downplay the inherently disruptive nature of convergence on the medium of radio itself and in many respects foreclosed avenues of innovation by which radio might have found more secure footing in our increasingly convergent media environment. Furthermore, as HD Radio was deployed in the real world, the system's inability to adequately address these conditions confused and sullied several important constituencies on the viability of digital broadcasting itself.

The FCC has been remarkably aloof regarding the development and promulgation of digital radio. The decision to adopt the HD system was made with no independent analysis, with a willful disregard of direct and meaningful public input, and with the use of economic metrics above all else to justify the industry's chosen technology. The FCC's reliance on the National Radio Systems Committee to provide the technical rationales for HD Radio vividly illustrates the moral bankruptcy of such decision making. Ostensibly, the NRSC represents a coalition of industry actors that works on a quasi-consensus model to settle on technological standards that are amenable to all parties and, in theory, advance radio's inherent utility. In the case of HD, this process devolved into folly: the membership of key NRSC subcommittees was hijacked during the 1990s by HD proponents that would not countenance any resistance to their system. As a result, constitutive NRSC evaluations of HD Radio were trumpeted as robust when they were actually woefully incomplete; some of the committee's most important decisions were effectively railroaded through, as those critical of HD were badgered into registering abstentions instead of opposition in order to keep the appearance of consensus afloat; and any hope of comparatively evaluating digital radio technologies failed when key NRSC members threatened to withhold their participation or filibuster discussions.

By the end of the decade, those driving the NRSC's digital radio work were wholly aligned with HD Radio's proprietor, iBiquity Digital Corporation. Since then, the NRSC has been nothing more than a rubber-stamp in the ongoing development of the HD standard, oftentimes passing off iBiquity-conducted—and funded—research as its own. The importance of broadcasters' control over the NRSC's digital radio deliberations cannot be overstated: this work effectively constrained the FCC's policymaking options regarding digital radio to the HD paradigm, and by the time the

FCC began formal rulemaking procedures any meaningful opportunity to critique, much less challenge, this paradigm had been effectively foreclosed.

Dallas Smythe's concern that "market forces" would come to dominate the "allocation of the spectrum as a whole" is embodied in the operative policy rationales of digital radio in the United States.[1] A spectrum-subsidy remains at the core of broadcasting, and HD Radio sweetens this subsidy in two new ways: by increasing the private control of available radio spectrum and providing the iBiquity cartel with the power to govern the functionality of *all* radio stations that operate in the digital domain. These unfortunate developments stem directly from the FCC's shedding of its ability to make empirically sound judgments in favor of faith-based regulation rooted in industry-provided datasets. This behavior is increasingly troublesome in a time when scientific impartiality and engagement, especially given the pace of convergence itself, is more necessary then ever. In the case of HD Radio, regulators seemed to decide early on that thinking outside the industry-defined box of what radio's future could be somehow threatened their own established policy practices.[2] If true, then the dilemma of digital radio poli-cymaking in the United States has significance beyond technological igno-rance or the simple favoritism of one constituency over another; it opens up for contestation the core ideology of the policymaking process and its outcomes.

The overarchingly neoliberal tenor of U.S. media policymaking has been strongly amplified by the Telecommunications Act of 1996. Were it not for the frenetic consolidation of radio station ownership during the critical years of HD Radio's development, and had the FCC actually taken the time and effort to seriously consider the system's fundamental detriments, the tech-nology just might have withered and died on the vine. Instead, the heavy-weights of the radio industry framed the adoption of HD Radio as necessary to "compete" with vaguely defined digital audio delivery services. From the FCC's perspective, the main selling point of HD was that it claimed the sup-port of those who controlled the majority of radio industry revenue. The policymaking calculus was clear—dollars were votes in determining just who represented "the public interest" in radio's digital transition. But this perspective omitted an important variable: the ownership of radio stations themselves, the vast majority of which reside outside the iBiquity cartel.

Thus, formative policy regarding radio's digital future was promulgated with casual heedlessness about what real-world effects digitalization would have on the radio industry as a whole, much less the medium itself. By the time HD Radio wended its way through an FCC already overburdened by its other duties as executor of the Telecom Act, the agency had essentially abdicated the regulation of digital radio to the most politically and eco-nomically powerful coalition of broadcasters. In this respect, the triumph of the private sector in radio's digital transition represents one of the Telecom Act's most impressive achievements with regard to cementing neoliberal principles at the heart of digital communications policy, and the saga of HD Radio is itself a capstone in the long and unfortunate history of the FCC as a captured regulator.

Examining the dramaturgy of digital radio policy development goes a long way toward explaining the convoluted and haphazard nature of the policymaking process. It draws attention to the arguments each constituency advanced (and how they advanced them), provides a mechanism by which they can be interrogated for substance, and allows the critical examination of policy outcomes. When conflicts arise between the operative rationales of contemporary digital radio policy and their practical application, it begs the question whether such regulation is actually constructive, or even rational. This question takes on new importance when prominent counternarratives contradict the stated rationales of policymaking, as was the case with the vibrant and feisty dialogue surrounding HD Radio that occurred both in the FCC record and trade press.

In simple terms, iBiquity Digital Corporation was founded to consolidate the core development of HD Radio within the broadcast industry itself and to provide a semblance of market consensus around the system that would convince regulators to sanction its proliferation. In this context, the technical controversies of HD were actually more of a sideshow than anything else. The exercise of convincing the FCC to adopt HD Radio was a well coordinated and intense lobbying effort, led primarily by iBiquity, its broadcast-investors, and the National Association of Broadcasters, and not without its fair share of deception and duplicity. This campaign functionally overwhelmed the FCC's pitiful resources and pressured the agency to move forward on digital radio before many of its fundamental attributes were even fully defined or understood. This is clear in the FCC's own formative rulemakings, in which caveats and prognostication outnumber substantive declarations about HD Radio's actual viability. Instead of providing a solid and defensible regulatory foundation for radio's digital transition, they set dubious precedents that have only complicated matters.

The rank hypocrisy of HD proponents in the technical vein becomes clear when juxtaposed with their rationales for opposing the promulgation of a low-power FM (LPFM) community radio service at the turn of the century. LPFM and HD Radio both represent an increased use of the FM spectrum by equivalent amounts (200 KHz per station). Broadcast incumbents opposed the creation of LPFM on the specious assertion that it amounted to "shoehorning" new spectrum allocations onto an already-crowded band in a manner that would open up the FM dial to destructive interference—even though an LPFM station is limited to a paltry 100 watts or less of broadcast power. However, the imposition of HD Radio—which required the FCC to fundamentally redefine the concepts of "channel" and "interference" on both the AM and FM dials—was contextualized as a negligible compromise to the integrity of the radio spectrum in the service of "progress."

The fruits of such tortuous regulation are difficult to swallow. Yet of HD Radio's three fundamental detriments, its proprietary nature has been its biggest albatross. The digital conversion rate of stations is pathetic; receivers are few and far-between; innovation within the HD Radio space is greatly hampered; and the failure of iBiquity's core business model has left the company without an independent and sustaining source of revenue. Wobbling along on infusions of radically devalued radio company stock and venture capital cannot continue indefinitely. The irony is that, in many respects, this outcome is precisely what HD proponents hoped for: radio's digital transition may be in a state of malaise, but it's *their* malaise. The best they can seem to manage in terms of tangible support for their chosen technology is a commitment to remedial fixes, furtive development efforts in hopes of finding some application for HD that will provide it with a semblance of tangible value (efforts which, on many levels, appear to be working at cross-purposes), and lukewarm promotional efforts that require little meaningful fiscal commitment. More than a dozen years on from its public introduction, this is a sorry state of affairs for a technology whose implications were promised to be revolutionary. Instead, much like a gambling addict who holds out hope that the next hand will turn their luck around, the iBiquity cartel is stuck at a game of its own making with no recourse but to keep doubling down.

Among the broadcast incumbency, although commercial conglomerates and the NAB did most of the heavy lifting, the critical interventions of public broadcasters in the development of HD Radio were vital to the "success" of this endeavor. Without them, there is a very good chance that radio's digital transition might have turned out quite differently. Robert McChesney asserts that "all broadcast and communication policy debates have been predicated upon the notion that the needs of the private sector come first, and that these are largely compatible with the public interest"—a perspective that has effectively marginalized noncommercial broadcasters in the United States since 1935.[3] Coupled with the fact that National Public Radio was an afterthought to the Public Broadcasting Act of 1967, it has constantly pursued "a niche on the margins, where [it] would not threaten the existing or potential profitability of commercial interests."[4] However, HD Radio represents the first time public broadcasters have actively promoted their future in such close consort with the those to whom they are supposed to be alternatives, if only to preserve a semblance of agency in the process of the digital transition. The fact that the nation's largest commercial and noncommercial broadcasters both backed HD Radio legitimized its adoption in the minds of regulators.[5]

Furthermore, NPR has been one of HD Radio's primary innovators (developing such features as multicasting and conditional access) and produced important technical data for FCC consumption that provided a patina of legitimacy to the technology that otherwise would not have existed. Although seemingly more astute than commercial broadcasters about the implications of convergence, NPR also did not recognize the inherent insufficiency of simply digitizing terrestrial radio's existing broadcast platform.

Finally, public broadcasters have spent significant public funds on HD Radio. Over the years, they received tens of millions of dollars in digital conversion subsidies, the lion's share of which came during a critical time in iBiquity's corporate life when other potential funding sources were drying up. The money public broadcasters have additionally spent on research and development, efforts such as providing multicast streams to NPR affiliates, and general lobbying and promotional activities is incalculable but significant. Even so, like the majority of commercial broadcasters who have opted out of the iBiquity cartel, noncommercial broadcasters by and large have not followed NPR's lead.

On balance, the behavior of public broadcasters with regard to radio's digital transition is downright shameful: NPR and many of its larger affiliate stations and networks were driven more by historical fears about potential shifts in the political economy of any new digital radio landscape than they were by true forward-thinking about digital radio's inherent potential to transform the medium. When public broadcasters made the strategic decision in the early 1990s to follow the lead of commercial broadcasters on digital radio development, they effectively became part of the problem, both from a technical and regulatory perspective. What is particularly contemptible is the fact that public broadcasters were cognizant of HD Radio's fundamental detriments from the outset and have remained sensitive to them over intervening years. But when these detriments manifested themselves in ways that could not be ignored, such as the controversy over raising the power of FM-HD sidebands, public broadcasters may have found themselves in principled opposition to the prevailing regulatory momentum but had cast away any meaningful agency many years earlier. In the end, the U.S. digital radio transition has actually crystallized public broadcasters' subservience to the corporate-centric regulatory paradigm in a manner not seen since the creation of the FCC itself.

The dramaturgy of HD proponents only represents half the story. The sheer volume of critical comment about the system found both in policy proceedings and the trade press—emanating from knowledgeable, well-respected industry participants and committed radio listeners—represents a stunning indictment of HD Radio's qualitative potential and vividly illustrates the captured nature of the FCC on fundamental questions of spectrum allocation, access to the airwaves, or any meaningful homage to conceptions of the public interest. As J. H. Snider noted in Chapter 5, the sheer political beauty of the HD Radio policymaking process was to force constitutive choices about the technology concurrent with FCC deliberation on the creation of the LPFM radio service. While those who would later sow the seeds of the modern media reform movement advocated for crumbs of analog spectrum under highly restrictive conditions, incumbent broadcasters were constructing a digital future for radio that would, at best, marginalize these new stations and, at worst, silence them.

The failure to engage in the early debates over digital radio's development was a tragic mistake on the part of modern media reformers. The data (or lack thereof) regarding HD Radio's fundamental viability should have been red flags to any public interest advocate thinking about the future of the medium; it provided the impetus for this book. Had an informed and engaged public been mobilized around the issue of digital radio when its key regulatory principles were being constructed, it—coordinated with the open reticence of independent broadcasters—might have provided an opportunity for coalition-building that could have forced a more comprehensive and honest consideration of digital radio itself. In retrospect, LPFM turned out to be an inopportune distraction that diverted the potential for organized opposition to the fundamental re-provisioning of spectrum that HD Radio represents.

By the time this opposition appeared, the operative rationales that digital radio was meant to address, as proffered by conglomerates and public broadcasters, were too ingrained in the regulatory paradigm to challenge effectively. It is remarkable that the lack of bona fide consensus around the viability of HD Radio was clear in the trade press long before the actual implementation of digital broadcasting began, and went unrecognized for what it portended regarding the sordid outcomes of the policymaking process. Trade press criticism of HD Radio, and its growth over time, vividly illustrates the detachment from reality that unfortunately pervades much of modern media policymaking.

POTENTIAL POLICY "RESOLUTIONS"

The lingering uncertainty over the fate of radio's digital future cannot last forever. A major question facing those involved in the digital transition is whether future investment in HD Radio can transform the system into a useful bridge to an increasingly convergent media environment. The system's deficiencies, coupled with the fact that its primary "innovations" merely mimic what other digital audio delivery platforms are already capable of, bear due consideration, for the utility of HD as a "blocking move" to prevent competition in the radio space is diminishing every day. This may ultimately force policymakers into more radical action that could very well open up the U.S. digital radio transition to a fundamental re-think.

Guaranteeing HD Radio's survival ultimately means forcing an analog/digital radio switchover. This would not be as simple as cribbing from the U.S. digital television transition: that process was written into the 1996 Telecom Act, which mandated a specific analog-cutoff deadline and provided upwards of $60 billion worth of new spectrum on which to accomplish it, as well as government subsidies to television viewers for their own conversion equipment so that receivers would not be obsoleted.[6] Furthermore, the DTV standard was the product of true consensus between broadcasters, electronics manufacturers, and telecommunications companies, and has been adopted by countries other than the United States.

On the other hand, HD Radio lacks the consensus among all of the constituents necessary to make a digital transition successful, due in large part to the opacity of the system's development. Its proprietary nature is precisely why the FCC made station conversion voluntary, which militates against any direct government subsidy. What's more, the system has no significant purchase elsewhere in the industrialized world, the majority of which has instead decided to either adopt other digital radio technologies or forego a transition in the near term. Nor is the transition to HD radio similar to the adoption of analog FM broadcasting, a common claim of HD proponents. FM radio was a true qualitative improvement on existing broadcast technologies, had listener interest from the outset, and (like DTV) had new spectrum with which to work; in contrast, HD is compromised by sharing spectrum with analog services, does not inherently provide new programming opportunities, and has little measurable listener interest.[7]

Any digital radio transition will ultimately involve an FCC mandate to cease analog operations, and the DTV experience suggests that this process will not be timely or orderly. Without a mandate, it's unlikely that many broadcasters will voluntarily adopt the all-digital mode of HD Radio, and the consumer electronics industry, whose clout is on the rise relative to broadcasters in Washington, D.C., has resisted and will continue to resist any digital receiver mandate attempt vigorously. Therefore, if the all-digital mode of HD Radio is never realized, the likelihood of its eventual demise increases, at which time the policy and practical objectives of digital radio will most definitely be reassessed. This may involve the exploration of other digital radio technologies, or perhaps the wholesale conversion of radio broadcast spectrum to other purposes, as regulators have a new hunger for spectrum by which to promote the wireless provision of broadband Internet access.

The U.S. radio industry and FCC are loath to acknowledge that other digital radio technologies exist that may provide a more viable path to digitalization than HD Radio. For example, Eureka 147 DAB is finding traction in many European countries. But it took more than 25 years to reach this point, due to regulatory fragmentation between and within various nation-states, coupled with a variety of technical hiccups, including disruptive innovations in the Eureka suite itself. Although digital radio broadcast listening remains a minority phenomenon in Europe and elsewhere where DAB has been deployed, broadcasters and consumer electronics manufacturers are finally working together to make sure all the components for a viable digital radio transition exist. Coupled with the active engagement of regulators, who are not nearly as cowed by industry forces as they are in the United States, there is at least enough coherency to Eureka-driven transition efforts that several countries are now mulling over a date-certain analog radio switchoff.[8] However, the spectrum on which Eureka 147 works is not available in the United States, nor is there any desire among broadcasters to move to a new-spectrum digital solution.

The newest digital radio broadcast system, Digital Radio Mondiale (DRM), does have promise but has been hobbled by being a latecomer. DRM's development began in March 1998, when representatives of research institutes, equipment manufacturers, and broadcasters from more than 30 countries met in China under the banner of the Digital Radio Mondiale Consortium to develop a digital standard specifically for the AM and short-wave bands. DRM wholly displaces analog radio signals, and the system is based on an open-source paradigm, eliminating any proprietary impediments to innovation and proliferation.[9] By 2001, a mobile DRM receiver was demonstrated at Germany's largest consumer electronics show, and in 2002 the International Telecommunication Union endorsed DRM for use on the AM and shortwave bands.[10]

The BBC, Voice of America, Deutsche Welle, DeutschlandRadio, Radio Canada International, Radio Netherlands, and Swedish Radio launched DRM shortwave broadcasts in 2003, targeting all continents on the planet except Antarctica.[11] Regulators in several countries saw the initial potential of DRM as a "drop-in replacement for existing . . . allocations," and also thought it might be useful for jump-starting stalled DAB system development by heightening consumer interest in digital radio more generally.[12] Between 2005 and 2008, more than 50 AM and shortwave broadcasters around the world adopted DRM technology, producing some 350 program-hours per day in the mode, and several countries began test programs to evaluate the possibilities of widespread DRM deployment.[13] H. Donald Messer, director of the Spectrum Management Division of the U.S. International Broadcasting Bureau, resigned his post in 2005 to work full-time with the DRM Consortium.[14] That same year, the consortium announced plans to develop an FM variant, known as DRM+, which was subsequently certified for global deployment in 2007.[15]

As the newest digital radio technology, the developers of DRM had the advantage of watching and learning from the pitfalls encountered by Eureka 147 and HD Radio. It enjoys an impressive international coalition of support that made the important constitutive decision to develop an open standard. The DRM Consortium has also pursued strategic partnerships with WorldDMB, including receiver interoperability, and often positions itself as a companion or supplement to Eureka 147-based technologies.[16] Qualitatively, DRM offers a perceptible increase in audio fidelity over analog radio broadcasts, and allows multicasting and datacasting on every band. However, the disruptive potential of entirely replacing incumbent analog signals has held back DRM's unqualified endorsement by any broadcaster or regulator.[17] The technology's uptake has also been stymied by a lack of receivers: Sony Corp., a member of the DRM Consortium, has declined to actually manufacture any DRM-compatible gear.[18] This reticence may change over the next decade as more countries expand their use of DRM; for example, India is considered a "sleeping tiger" of digital radio, and its public service broadcaster has committed to a DRM transition on AM and shortwave

during this decade.[19] Domestically, at least one manufacturer already makes
a transmitter that's cross-compatible with DRM and AM-HD Radio, so
the technology is well within the reach of U.S. broadcasters—provided they
and the FCC were actually amenable to the consideration of alternative
technologies.

That said, most broadcasters and regulators are seeking an all-in-one digital radio standard to adopt, and present commitments to DRM+ are at best
tentative. It is easy for DRM to colonize the shortwave band, because it's
the only digital radio technology available there, and as a niche broadcasting service what happens on shortwave is unlikely to change the trajectory
of ongoing digital radio transition attempts. However, the fact that many
developing nations, including three of the four BRIC countries, seem serious
about adopting DRM certainly portends promise.

Perhaps the way to avoid the problems of a digital radio transition is not
to have one at all—at least not until extant technologies have more time to
mature. This may seem like heresy to many modern broadcasters, but the
case of Canada begs to differ: it is possible to experiment with and fail at
a digital radio transition and survive. In 1995, Canada formally endorsed
the Eureka 147 DAB system and organized testing began immediately.[20]
As regulators there solicited multiplex licensees, commercial broadcasters began preparing to join in the provision of DAB service.[21] However,
as the costs of building the new transmission infrastructure became better
understood, commercial interest in DAB cooled. Some broadcasters began
to openly question whether promises of increased audio fidelity would be
enough to sell the technology to a disinterested public.[22] DAB receivers were
expected to hit the Canadian market in "mid to late 1997" and that year the
Canadian Broadcasting Corporation and many commercial broadcasters
launched DAB multiplexes in Canada's largest metropolitan areas.[23] General Motors of Canada subsequently announced that it would include DAB
receivers as standard equipment across its entire line of vehicles, and Ford
expressed similar enthusiasm for the technology.[24]

The uptake of digital radio in Canada did not materialize. By 2002, there
were 57 DAB stations on the air, reaching just 35% of the population, and
receivers remained scarce and expensive.[25] In 2004, General Motors of Canada, citing "difficulties . . . over supply of equipment," rescinded its commitment to include DAB receivers in its vehicles. Listener reaction to digital
radio was underwhelming: complaints circulated about the audio quality,
and DAB signals were not as robust as their analog counterparts. Ultimately,
commercial broadcasters used the technology primarily to simulcast their
analog stations.[26] The Canadian Association of Broadcasters—the trade
group for the country's commercial broadcasting sector—argued in 2005
that it was "simply not realistic" to assume that DAB would ultimately
replace analog radio service.[27] The following year, 11 multiplexes went
silent; those that remained reached a potential audience of just 11 million

listeners, and radio audience measurement services stopped quantifying digital radio listenership altogether.[28]

Acquiescing to these negative developments, Canadian regulators opened up radio's digital transition to Eureka alternatives.[29] This resulted in a brief affair with HD Radio, but after a year of testing by public and commercial broadcasters, the CBC and Radio Canada concluded that they would "make no further investments in [HD Radio] until the interest of other Canadian broadcasters is gauged and while it monitors the rollout of data services and applications in the United States."[30] Station owners appeared to be more concerned about the potential of HD-related interference from U.S. stations than they were about adopting the technology themselves.[31]

By 2010, Canada's digital radio transition had effectively disintegrated. As broadcasters openly declared the Eureka system to be in "limbo" and "peril," the CBC shuttered four DAB channels in Montreal. This was interpreted as part of an industry-wide abandonment of the platform.[32] At the same time, the Canadian Association of Broadcasters disbanded.[33] Suddenly there was no coherent broadcast constituency left to advance the cause of digital radio in Canada. Regulators have since proposed reallocating DAB spectrum for fixed and mobile wireless devices, and ultimately they would like to see broadcasters develop a digital platform that complements existing analog broadcasting services. At present, Canadian broadcasters are wholly unprepared to assume such a task.[34] Yet, the demise of the Canadian radio industry has not come to pass; if anything, the route of non-participation enables Canada's broadcasters and regulators time to more properly consider radio's ultimate place in a convergent media environment, making future investments in a transition much more fruitful. We must also not forget that a large segment of the world's population lives in regions, such as South America and Africa, where analog radio broadcasting remains an integral part of the media environment and no pressing plans exist to force a digital transition.

One of the reasons why radio's digital dilemma is so pressing is because convergence is also having a significant effect on thinking about just what the public airwaves might be used for. Digital broadcasters have already employed their channel allocations for non-broadcast purposes. For example, SiriusXM satellite radio provides a Federal Aviation Administration-approved service that transmits real-time weather information for use by pilots. In the realm of digital television, there has been much experimentation with alternate uses of spectrum: Clear Channel tried to provide downstream Internet access via DTV; Disney once leased portions of local DTV channels to transmit movies-on-demand to subscribers, as did a league of broadcast and companies that also provided music, games, and software; and News Corporation attempted to undercut the bottom end of cable TV's subscriber base by feeding 12 encrypted basic cable channels over the air for just $20 per month.[35] None of these ventures gained traction because existing wired broadband networks provide such services more efficiently than a DTV conduit ever could. In the context of radio, the problem is that a single AM or FM channel, even fully digitized, does not contain enough capacity to provide broadband-level services. However, *cumulatively* organizing radio bandwidth is theoretically possible, and such aggregation would radically change the identity of radio in its own right. The fact that economic incentives exist for incumbent broadcasters to reemploy their spectrum for purposes other than broadcasting deserves greater scrutiny—as does the U.S. government's increasingly open discussion of such concepts.

In 2010, the FCC released a National Broadband Plan that called for the repurposement of several hundred megahertz of spectrum for wireless broadband provision.[36] The agency is looking far and wide for such spectrum, including among incumbent users of the airwaves. That November, the FCC promulgated a Notice of Proposed Rulemaking seeking to reclaim some 120 MHz of DTV spectrum for broadband Internet access.[37] Considering that a single DTV channel covers approximately 6 MHz, the FCC seeks to repurpose 20 DTV channels' worth of spectrum in all.

In 2012, the FCC unveiled an "incentive auction" proposal by which the government will buy back this spectrum from broadcasters.[38] A group named the Expanding Opportunities for Broadcasters Coalition has since been formed to represent the television industry in behind-the-scenes negotiations with the FCC over terms of the auction itself; the coalition now represents "70 stations, mostly in the larger markets where the FCC is most in need of spectrum to reclaim for wireless," and while its membership is confidential—because "it would not be in a station owner's interest to signal to competitors or its own staffers that it might be giving up spectrum"—the coalition's executive director is a former chief lobbyist for Disney and Fox.[39]

AM radio spectrum is not a good candidate for such repurposement, given the paucity of bandwidth occupied by the entire band and the propagation characteristics of AM signals. In addition, as noted in Chapter 7, AM broadcasters are using the back door of FM translator stations to simulcast their signals in an environment where HD Radio has some recognizable functionality. This suggests a shift toward FM as the singular U.S. terrestrial digital radio platform of the future is already underway. On the other hand, the FM dial occupies a cumulative 20 MHz of spectrum, and its properties are much better suited for wireless broadband provision: the line-of-sight and building-penetration capabilities of FM radio signals are similar to those found in the VHF band of television.[40] Should HD Radio fail to guarantee the broadcast industry its spectrum subsidy, the FCC might very well open the door to refarming FM spectrum as part of its campaign to promote more wireless broadband. In 2011, *Radio World* pointedly asked NAB president Gordon Smith whether there was "any immediate threat" to the radio spectrum; he responded, "Not immediate, but if they can do it to your neighbor [broadcast TV], they can do it to you eventually."[41]

Although spectrum repurposement is an unlikely outcome of current digital radio policymaking initiatives, it may be the one most worthy of further exploration. In the words of Michael McCauley, it is the duty of media scholars engaged in policy studies to not let "vested interests . . . define their questions or their area of inquiry in such a way as to see 'realistic' policy proposals as those which will be acceptable to the vested interests in the short term and without a struggle. It is always dangerous in such an undertaking to try to guess what will be acceptable and it can act as a distraction from the very real difficulties of working out and then fighting for what you believe to be true."[42]

RADIO'S EVOLVING IDENTITY

It seems increasingly likely that the success or failure of HD Radio ultimately lies in the hands of the consumer electronics industry. Broadcasters sullied their enthusiasm for the system by trying to coerce its adoption, first by dictating the system's development process and later by convoluted attempts at backdoor policymaking. As a result, electronics manufacturers opted out of development efforts long before iBiquity Digital Corporation was even founded, and have effectively ignored the technology ever since. Lacking their active and widespread support, a primary ingredient necessary for HD Radio's successful uptake has never materialized. Broadcasters continue to needle the consumer electronics industry toward a wider embrace of terrestrial radio through ongoing lobbying and public relations campaigns that attempt to position FM reception as a must-have in portable wireless devices. But the industry is also increasingly exploring the carrot-approach, which at this late stage involves buying access to selected device-markets outright.

In contrast, the Internet Protocol (IP)-based streaming element of new radio services provides functionality within existing digital media infrastructures for which many products are already designed and with a hugely attractive cost-benefit ratio. It is no wonder why consumer electronics manufacturers have effectively ostracized HD; now that other variants of "radio" exist, and their methods of delivery are beginning to approach levels of accessibility competitive with terrestrial broadcasting, HD Radio's place in the universe of digital media devices is a marginal one.

Taking the dashboard as an example, what was once a listening space traditionally dominated by broadcasting is now wholly up for grabs. Satellite radio subsidized its position in the car by giving auto manufacturers a cut from listener subscription fees in order to entice them to add satellite reception capability to vehicles, thereby making the sale of each vehicle more profitable for its manufacturer. Terrestrial radio broadcasters simply can't afford to pursue a similar strategy. Furthermore, vehicle manufacturers are creating in-dash "infotainment" centers that allow drivers to link other digital media devices into them. Current iterations of this technology, which tether smartphones to the dashboard or, in some vehicles, provide direct mobile wireless broadband access, increase the menu of available "radio" choices, further fragmenting listenership. It is clear that vehicle manufacturers, and the consumer electronics companies that supply them, are making these investments on the assumption that "radio" is no longer limited to the AM or FM broadcast bands. This mentality exists across every electronics platform where radio once enjoyed a place of primacy.

Among listeners, the medium of radio itself has already been redefined to include subscription-based satellite broadcast services as well as a plethora of customizable Internet-based audio content providers such as Pandora, LastFM, Slacker, Spotify, and the larger world of podcasting. In 2013, Google and Apple both entered the online "radio" universe with their own streaming services (one might argue that Google has been in this space for years, as YouTube is used as a music-discovery tool in its own right). Many incumbent broadcasters are cognizant of this competition and now also provide their content online; Clear Channel is the most ambitious in this regard, aggregating the streams of its radio stations (as well as many other commercial and noncommercial broadcasters) via iHeartRadio, while NPR has secured distribution of its programming over the SiriusXM satellite radio network and is hard at work developing mobile applications for its content. As broadcasters move to these platforms to compete directly with new digital radio services, their distinctiveness relative to these new services is blurred, which further complicates the public identity of the medium in the twenty-first century.

However, many new digital "radio" services remove the human element from the provision of content, relying instead on algorithms designed to "learn" a listener's tastes and suggest programming for them; this is why these services don't (yet) call themselves broadcasters. However, when incumbent broadcasters marginalized the human element of their programming in favor of a more highly consolidated and automated industry, they forfeited an element of distinctiveness that has not gone unnoticed by the listening public. Listeners have repeatedly complained about the lack of program diversity and localism in today's radio environment. It is rare to find such a richly cogent documentation of this sentiment as exists in the FCC's rulemakings on digital radio, heretofore considered a point of policy esoterica. It suggests that the public still finds value in the act of broadcasting, which cannot be wholly replicated or displaced by newer forms of digital "radio."

Legacy media systems have a long history of difficulty navigating disruptive change. Early conceptions for use of the radio spectrum assumed it would simply extend the practice of telegraphy. Gone unacknowledged by many were innovations during the early twentieth century that used radio for means other than the point-to-point relay of Morse-coded messages. Until very recently, broadcasting defined the medium of radio.[43] In the same regard, broadcasters first cast convergence as a phenomenon that would simply extend the practice of radio; now that radio itself is being redefined, they have resisted understanding and accommodating the inevitable changes. How radio stations might be able to profitably port the practice of broadcasting into a convergent media environment remains to be substantively explored.

That said, it is dangerous to categorize the growth of other "radio" services as threats that may entirely subsume the identity of the medium as we've known it. SiriusXM, the monopoly provider of U.S. satellite radio, has a subscriber base of 24 million listeners after more than ten years on the air, and since the company has already achieved profitability, increasing its subscriber base is but one of several options that may drive its future growth.[44] Ubiquitous service is not the ultimate goal of satellite radio, whereas the development of terrestrial radio was designed (and still mostly functions) with that goal in mind. Furthermore, ongoing industry and policy developments regarding access to the Internet itself impart a significant element of uncertainty to the growth and maturity of streaming audio services. Companies that dominate Internet access and the provision of online content are pushing for regulatory conditions that would give them increasing power over the basic functionality of convergent media platforms.[45] For example, where unlimited data consumption used to be the norm with regard to Internet access, usage caps are now being imposed by broadband service providers that could significantly limit a listener's ability to "tune in" radio online.[46] The ongoing commodification of broadband cannot help but have detrimental impacts on the communicative potential of the Internet itself.[47]

There is little doubt that radio will one day be all-digital in nature, but the mechanisms by which radio broadcasters will evolve in that direction remain relatively underdeveloped. Unfortunately, the attempted digitalization of existing radio stations is but the first step in an arduous process that has years left to play out. The longer that broadcasting in the United States remains marginalized by its own design, the more likely it is that the crisis of identity among those that have historically claimed ownership of "radio" will be exacerbated. The solution to this quandary will not be found in a technological panacea. The Internet of the present shows us a glimpse of what a future universal media infrastructure may look like.[48] But at the same time, the Internet itself is a network of networks whose practical makeup is more an amalgam than a unified entity, and whose regulation also remains in an uncomfortable state of flux. Despite this, the fact that radio now exists on-air and online, via wired and wireless conduits, may diminish the perceived necessity of the medium's incumbent transmission infrastructure, especially as listening continues to migrate away from the AM and FM dials themselves and toward new services that continue to transform the public's definition of "radio" itself.

In the midst of this great uncertainty, there still exists a strong sentiment within the public that believes the FCC's mandate to serve the "public interest, convenience and necessity" should have some literal meaning. In the time preceding the industry's consolidation, this was primarily expressed through concepts of localism, which helped to facilitate program diversity and strengthened the communicative act of broadcasting itself.[49] Some believe that, were the FCC to simply return to some core principle of local-ism in the regulation of radio, its digital dilemma could be addressed pro-ductively and without danger to the integrity of the medium;[50] at least a few FCC commissioners and staff appear inclined to agree.[51] However, the con-cretization of neoliberal ideology within media policy circles in the wake of the Telecom Act, coupled with the disruptive influence of new digital media technologies which contest the identity of "radio," means that a return to localism alone is not likely to alter a system of broadcast policy which has a built-in bias away from this principle.[52] This raises the uncomfortable question of whether a "a viable broadcast reform movement" to "enhance radio's public service function in the digital era"[53] can achieve meaningful currency in a convergent media environment, where the notion of broad-casting itself is being openly challenged.[54]

So long as spectrum-incumbents firmly control both access to and the means of digital development of the airwaves, the likelihood of proactively influencing the evolutionary trajectory of digital radio through conventional means is greatly diminished.[55] A functional inability to simultaneously address media policy through the lens of convergence and in the context of meaningful public interest standards only exacerbates this problem.[56] The increasing propertization of digital media itself is clearly a step in the wrong direction, and suggests that "support for a neo-liberal ideology that views government restrictions as political impediments to the success attainable via the free market" will continue to control contemporary media policy-making, despite its well-documented pitfalls.[57]

Key to claiming agency in the regulatory development of new technologies is to act "before an unplanned commercial system becomes entrenched," so that its social value can be assessed.[58] If a major function of communications policymaking is to understand the "unanticipated or unintended effects" of new technologies "so that we may avoid or minimize the undesirable ones,"[59] the saga of HD Radio on a regulatory level is one of abject failure. However, this does not mean that all is lost. Tim Wu has suggested that new media technologies go through a "cycle" of openness and closure: openness spurs innovation and uptake, while closure is precipitated by attempts to turn a technology toward a for-profit paradigm. "History also shows that whatever has been closed too long is ripe for ingenuity's assault, in time a closed industry can be opened anew, giving way to all sorts of technical pos-sibilities and expressive uses for the medium before the effort to close the system likewise begins again."[60]

In the context of digital radio's development in the United States, the phase of openness never really happened, for HD Radio was predisposed to closure, and its present malaise suggests that a turn toward openness may be in order. Yet throughout the history of broadcasting in the United States, regulators have failed "to affirm a considered vision of what broadcast-ing should be, only following and accommodating the evolution of busi-ness models."[61] Will radio's digital dilemma eventually force the breaking of that mold? Those who still care about the future of the medium need to engage in structural broadcast reform efforts—both analog and digital—and aspire to a more comprehensive understanding about the implications of convergent media policy more broadly. The unsettlement surrounding radio's digital transition does hold out some hope that there is still potential to promote meaningful change to the present trajectory, and communication scholars in particular have an important role to play, especially as it relates to "wield[ing] the weapons of intellect and scholarship that expose the workings of the corporate media system."[62]

Eric Klinenberg has asserted that media reformers can draw lessons from the environmental movement of the 1960s, which brought a network of disparate interests together in an effort to fundamentally reshape the world in which we live.[63] The impetus for interests to coalesce in the manner that Klinenberg suggests ultimately takes such a movement far beyond radio, with eyes firmly locked on the goal of influencing the pace and "rules" of convergence itself. Understanding and working to change the trajectory of radio's digital future would be a good object lesson from which to steepen the learning curve of contemporary media reform efforts. Under present circumstances, the convergence phenomenon will continue to "be a kind of kludge—a jerry-rigged relationship among different media technologies—rather than a fully integrated system. Right now, the cultural shifts, the legal battles, and the economic consolidations that are fueling media convergence are preceding shifts in the technological infrastructure. How those various transitions unfold will determine the balance of power in the next media era."[64] Recognizing and resisting the marketplace metaphor that now controls nearly all frames of acceptable discourse around media policy and practice is key to shaping a convergent media environment with maximum democratic potentiality. The story of radio's digital dilemma illustrates just how far away that future really is, and just how much work remains to be done if we truly desire to actualize it.

NOTES

1. Dallas W. Smythe, "Radio: Deregulation and the Relation of the Private and Public Sectors," *Journal of Communication* 32, no. 1 (Winter 1982): 200.
2. Ernest A. Hakanen, "On Autopilot Inside the Beltway: Organizational Failure, the Doctrine of Localism, and the Case of Digital Audio Broadcasting," *Telematics and Informatics*, 12, no. 1 (1995): 13.
3. Robert W. McChesney, *The Political Economy of Media: Enduring Issues, Emerging Dilemmas* (New York: Monthly Review Press, 2008): 236.
4. Ibid., 239.
5. The same unholy alliance also decimated LPFM: commercial broadcasters and the NAB cooked up the disingenuous technical argument in opposition to it, while NPR provided political legitimacy that carried the day.
6. For more on the DTV transition and its significant differences to the trajectory of radio's digital; transition, see J. H. Snider, *Speak Softly and Carry a Big Stick: How Local TV Broadcasters Exert Political Power* (Lincoln, NE: iUniverse, 2005).

7. Paul Thurst, "HD Radio 2010 = FM Radio 1950 (not)," *Engineering Radio*, September 24, 2010, http://www.engineeringradio.us/blog/2010/09/hd-radio-2010-fm-radio-1950-not/.
8. John Nathan Anderson, "Radio Broadcasting's Digital Dilemma," *Convergence: The International Journal of Research into New Media Technologies* 19, no. 2 (May 2013), 177–190.
9. See Marguerite Clark, "DAB Systems Are Now a Reality," *Radio World*, May 13, 1998, 41–42; and Peter Senger, "DRM Takes Aim at Digital AM," *Radio World*, June 10, 1998, 10.
10. See Leslie Stimson, "Mobile DRM DAB System Unveiled at IFA Show," *Radio World*, October 24, 2001, 6; Leslie Stimson, "DRM Demos Receiver at IBC," *Radio World*, October 9, 2002, 8; Leslie Stimson, "DRM Reaches for Launch," *Radio World*, May 21, 2003, 2; and Leslie Stimson, "DRM: Shortwave Wants to Sound Better, Too," *Radio World*, June 18, 2003, 3, 5.
11. "DRM Launches Big," *Radio World*, July 16, 2003, 7.
12. Lawrie Hallett, "DRM: What's It About?" *Radio World*, October 22, 2003, 1, 8, 10.
13. See Leslie Stimson, "Russia Chooses DRM for Digital Pilot," *Radio World*, February 1, 2004, 3; Leslie Stimson, "Chinese Radio Supports DRM," *Radio World*, May 5, 2004, 8; Daniel Mansergh, "DRM, Eureka Grow Worldwide," *Radio World*, July 1, 2004, 3, 8; James Careless, "CBC Enters Year 2 for DRM Tests," *Radio World*, December 1, 2004, 6; Jeff Cohen, "More DRM Radios Due in 2005," *Radio World*, January 19, 2005, 1, 6; "Macedonian AM Tries DRM," *Radio World*, February 2, 2005, 3; "DRM Tests Begin in Mexico," *Radio World*, March 30, 2005, 2, 7; Leslie Stimson, "DRM to Expand to FM, Combine Efforts With DAB," *Radio World*, April 12, 2005, 7; Lawrie Hallett, "Norway Looks at DRM Options," *Radio World*, September 16, 2005, 16; "DRM Eyes India for MW Test," *Radio World*, June 6, 2007, 2; Fabio Carera, "DRM Trial Under Way in Italy," *Radio World*, September 24, 2008, 18, 20; and "Newswatch," *Radio World*, December 17, 2008, 12.
14. Leslie Stimson, "Messer Leaves IBB for DRM," *Radio World*, August 3, 2005, 2, 6; and Jeff White, "Shortwave Awaits DRM in United States," *Radio World*, July 18, 2007, 5–6.
15. See Lawrie Hallett, "DRM Expands Into FM Band," *Radio World*, June 8, 2005, 8; Daniel Mansergh, "DRM Hopes for FM Standard in 2007," *Radio World*, September 13, 2006, 6, 8; and Daniel Mansergh, "DRM: Ready for Global Deployment," *Radio World*, July 18, 2007, 3.
16. T. Carter Ross, "Globally, Digital Radio Progresses," *Radio World*, June 5, 2013, 1.
17. Marko Ala-Fossi, "The Technological Landscape of Radio," in Brian O'Neill, Marko Ala-Fossi, Per Jauert, Stephen Lax, Lars Nyre, and Helen Shaw, eds, *Digital Radio in Europe: Technologies, Industries and Cultures* (Bristol: Intellect Ltd., 2010), 50.
18. Michael Dumiak, "Europe Dithers Over Digital Radio," *IEEE Spectrum* 43, no. 10 (October 2006), 20.
19. Ross, "Globally, Digital Radio Progresses," 10, 12.
20. See James Careless, "DAB Plan Enrages Canadian Broadcasters," *Radio World*, July 12, 1995, 11; Lynn Meadows, "AFCCE Hears Eureka 147," *Radio World*, July 26, 1995, 3; and Brian O'Neill, "Beyond Europe: Launching Digital Radio in Canada and Australia," in O'Neill et al., *Digital Radio in Europe*, 137–138.
21. James Careless, "Canada DAB on Track," *Radio World*, November 29, 1995, 1, 8.
22. James Careless, "Canadians Have Doubts About DAB Roll Out," *Radio World*, March 20, 1996, 14.

23. James Careless, "DAB Allocation Begins in Canada," *Radio World*, September 4, 1996, 8, 15.
24. See James Careless, "Ford Pays Attention to DAB in Canada," *Radio World*, December 11, 1996, 3; James Careless, "Canada Experiments With Digital-Only," *Radio World*, April 10, 2002, 28; and Brian O'Neill, "Digital Audio Broadcasting in Canada: Technology and Policy in the Transition to Digital Radio," *Canadian Journal of Communication* 32, no. 1 (2007): 77.
25. See James Careless, "DAB at the Border Line," *Radio World*, January 7, 1998, 1, 3; Skip Pizzi, "Rethinking DAB North of the Border," *Radio World*, August 1, 2004, 17–18; and O'Neill, "Beyond Europe," 140.
26. O'Neill, "Beyond Europe," 140–141.
27. O'Neill, "Digital Audio Broadcasting in Canada," 86.
28. See James Careless, "CAB: Digital Radio Needs Help," *Radio World*, September 13, 2006, 19; O'Neill, "Digital Audio Broadcasting in Canada," 77–80; and O'Neill, "Beyond Europe," 140.
29. O'Neill, "Digital Audio Broadcasting in Canada," 85.
30. See Wayne A. Stacey, "Canada Eyes IBOC Additions to DAB," *Radio World*, January 17, 2007, 23, 26; and Leslie Stimson, "Canada Allows IBOC," *Radio World*, August 1, 2007, 8.
31. Tom Vernon, "Here's What's on Their Minds," *Radio World*, March 25, 2009, 18, 20.
32. See Leslie Stimson, "Canadian Broadcasters Start Turning Off DAB Transmitters," *Radio World*, June 24, 2010, http://www.rwonline.com/article/canadian-broadcasters-start-turning-off-dab-transmitters/5195; and James Careless, "Canada in Digital Radio Limbo," *Radio World*, October 28, 2010, http://www.radioworld.com/article/canada-in-digital-radio-limbo/4078.
33. James Careless, "Canadian Lobby No More," *Radio World*, June 8, 2010, http://www.rwonline.com/article/canadian-lobby-no-more/3339.
34. See Grant Goddard, *D.A.B. Digital Radio: Licensed to Fail* (London: Radio Books, 2010); and O'Neill, "Beyond Europe," 143–145.
35. John Anderson, "Digital Radio in the United States: Privatization of the Public Airwaves?" *Southern Review* 39, no. 2 (2006): 18–19.
36. Federal Communications Commission, "Spectrum," in *National Broadband Plan: Connecting America,* April 21, 2010, http://www.broadband.gov/plan/5-spectrum/.
37. Federal Communications Commission, Notice of Proposed Rulemaking, *In the Matter of Innovation in the Broadcast Television Bands: Allocations, Channel Sharing and Improvements to VHF*, Docket No. ET 10–235, November 30, 2010, http://hraunfoss.fcc.gov/edocs_public/attachmatch/FCC-10-196A1.pdf.
38. Federal Communications Commission, Notice of Proposed Rulemaking, *In the Matter of Expanding the Economic and Innovation Opportunities of Spectrum Through Incentive Auctions*, FCC 12–118, Docket No. 12–268, September 28, 2012, http://hraunfoss.fcc.gov/edocs_public/attachmatch/FCC-12-118A1.pdf.
39. See John Eggerton, "Coalition Now Comprises 70 Station Members," *Broadcasting & Cable*, June 5, 2013, http://www.broadcastingcable.com/article/493878-Padden_Coalition_Now_Comprises_70_Station_Members.php, and *Expanding Opportunities for Broadcasters Coalition*, http://broadcastcoalition.org/.
40. In fact, VHF Channels 5 and 6 are contiguous to the FM radio band, spanning 76–88 MHz. In other countries, such as Japan, this spectrum is already used for FM radio broadcasting.
41. Paul McLane and Leslie Stimson, "Gordon Smith, Back on the Hill, Puts Engagement to the Test," *Radio World*, March 23, 2011, 6.

42. Nicholas Garnham, ed. Fred Inglis, *Capitalism and Communication: Global Culture and the Economics of Information* (London: Sage, 1990), 102.
43. Susan J. Douglas, *Inventing American Broadcasting, 1899–1922* (Baltimore: Johns Hopkins University Press, 1987), xv, xiv.
44. Trefis Team, "Can Sirius XM Tune In Big Subscriber Growth This Year?," *Forbes*, April 12, 2013, http://www.forbes.com/sites/greatspeculations/2013/04/12/can-sirius-xm-tune-in-big-subscriber-growth-this-year/.
45. See Federal Communications Commission, "FCC Acts to Preserve Internet Freedom and Openness," December 21, 2010, http://hraunfoss.fcc.gov/edocs_public/attachmatch/DOC-303745A1.pdf; Nate Anderson, "Verizon Sues FCC, Says 'Net Neutrality Lite' Laws are Illegal," *Ars Technica*, January 20, 2011, http://arstechnica.com/tech-policy/news/2011/01/verizon-sues-fcc-says-net-neutrality-lite-rules-illegal.ars; Susan P. Crawford, "The Looming Cable Monopoly," *Yale Law and Policy Review* 29 (December 16, 2010): 34–40; and Rob Frieden, "New, Old and Forgotten Frames in the Network Neutrality Debate," *TeleFrieden*, January 6, 2011, http://telefrieden.blogspot.com/2011/01/new-old-and-forgotten-frames-in-network.html.
46. See Larry Dignan, "Wireless Data Caps: Are Usage Based Pricing Schemes Here to Stay?" *ZDNet*, March 10, 2009, http://www.zdnet.com/blog/btl/wireless-data-caps-are-usage-based-pricing-schemes-here-to-stay/14097; Stacey Higginbotham, "T-Mobile Drops Cap on Mobile-Broadband Data Usage," *Gigaom* via *Businessweek*, April 27, 2010, http://www.businessweek.com/technology/content/apr2010/tc20100427_282228.htm; Esme Vos, "AT&T new wireless data plans impose data caps," *MuniWireless*, June 2, 2010, http://www.muniwireless.com/2010/06/02/att-new-wireless-data-plans-imposes-data-caps/; Mike Masnick, "Sprint Realizing That Data Caps Turn Customers Off," *Techdirt*, July 15, 2010, http://www.techdirt.com/blog/wireless/articles/20100715/16403510234.shtml; and Philip M. Dampier, *Stop The Cap!*, last modified June 20, 2013, http://stopthecap.com/.
47. For a comprehensive treatment of this particular dilemma, see Susan Crawford, *Captive Audience, The Telecom Industry and Monopoly Power in the New Gilded Age* (New Haven, CT: Yale University Press, 2013); and Robert W. McChesney, *Digital Disconnect: How Capitalism Is Turning the Internet Against Democracy* (New York: The New Press, 2013).
48. Tim Wu, *The Master Switch: The Rise and Fall of Information Technologies* (New York: Alfred A. Knopf, 2010), 256.
49. Michael P. McCauley, "Radio's Digital Future: Preserving the Public Interest in the Age of New Media," in Michele Hilmes and Jason Loviglio, eds, *Radio Reader: Essays in the Cultural History of Radio* (New York: Routledge, 2002), 506–507.
50. Hakanen, "On Autopilot Inside the Beltway," 19.
51. See Michael J. Copps, "Getting Media Right: A Call to Action," speech delivered at the Columbia University of School of Journalism, New York, December 2, 2010, http://hraunfoss.fcc.gov/edocs_public/attachmatch/DOC-303205A1.pdf; and Sherille Ismail, *Transformative Choices: A Review of 70 Years of FCC Decisions*, Federal Communications Working Paper, October 2010, http://hraunfoss.fcc.gov/edocs_public/attachmatch/DOC-302496A1.pdf, 25.
52. Patricia Aufderheide, "Shifting Policy Paradigms and the Public Interest in the U.S. Telecommunications Act of 1996," *The Communication Review* 2, no. 2 (1997): 261.
53. McCauley, "Radio's Digital Future," 525–526.
54. See Jo Tacchi, "The Need for Radio Theory in the Digital Age," *International Journal of Cultural Studies* 3, no. 2 (2000): 292–293; and O'Neill, "Digital Audio Broadcasting in Canada," 73–74.

55. Thomas Streeter, *Selling the Air: A Critique of the Policy of Commercial Broadcasting in the United States* (Chicago: University of Chicago Press, 1996), 194–195.
56. See Dan Schiller, *Digital Capitalism: Networking the Global Market System* (Cambridge, MA: The MIT Press, 1999), 209; and Jeff Chester, *Digital Destiny: New Media and the Future of Democracy* (New York: The New Press, 2007), 192–208.
57. Jonathan A. Obar, "Beyond Cynicism: A Review of the FCC's Reasoning for Modifying the Newspaper/Broadcast Cross-Ownership Rule," *Communication Law and Policy* 14, no. 4 (Autumn 2009): 484, footnote 24.
58. Robert W. McChesney, *Rich Media, Poor Democracy: Communication Politics in Dubious Times* (New York: The New Press, 2000): 126–127.
59. McChesney, *The Political Economy of Media*, 378.
60. Wu, 6.
61. Ibid., 85.
62. Larry Gross, "Fastening Our Seatbelts: Turning Crisis into Opportunity," *Journal of Communication* 62 (2012): 919–931.
63. Eric Klinenberg, *Fighting For Air: The Battle to Control America's Media* (New York: Metropolitan Books, 2007), 13.
64. Henry Jenkins, *Convergence Culture: Where Old and New Media Collide* (New York: New York University Press, 2006), 17.

Index